Nano-Optics and Nanophotonics

The Springer Series in Nano-Optics and Nanophotonics provides an expanding selection of research monographs in the area of nano-optics and nanophotonics, science- and technology-based on optical interactions of matter in the nanoscale and related topics of contemporary interest. With this broad coverage of topics, the series is of use to all research scientists, engineers and graduate students who need up-to-date reference books. The editors encourage prospective authors to correspond with them in advance of submitting a manuscript. Submission of manuscripts should be made to the editor-in-chief, one of the editors or to Springer.

More information about this series at http://www.springer.com/series/8765

Takashi Yatsui

Nanophotonic Chemical Reactions

New Photochemical Reactions and Their Applications

Springer

Takashi Yatsui
Department of Electrical
and Electronic Information Engineering
Toyohashi University of Technology
Toyohashi, Aichi, Japan

ISSN 2192-1970 ISSN 2192-1989 (electronic)
Nano-Optics and Nanophotonics
ISBN 978-3-030-42841-9 ISBN 978-3-030-42843-3 (eBook)
https://doi.org/10.1007/978-3-030-42843-3

This Springer imprint is published by the registered company Springer Nature Switzerland AG
The registered company address is: Gewerbestrasse 11, 6330 Cham, Switzerland

This book is dedicated to
late Prof. Katsuyuki Nobusada.

Preface

This book outlines "non-uniform" optical fields and their applications. The author has been working on nanophotonics for the quarter of a century. At the beginning of the work, the author has focused in improving the performance of the optical probes for ultra-high-density optical memory devices. Through the studies, the author has achieved high throughput probes to generate a strong optical near-field (ONF) from the propagating far-field light. Using such a technique, additional applications were developed such as nanoscale fabrication by scanning fiber probes and photolithography.

Since the scanning fiber probes are time-consuming, the author has shifted to the development of self-assembling techniques. In addition, many interesting phenomena have been observed using wavelengths longer than the absorption edge wavelength. These have been summarized in the book, *Nanophotonic Fabrication: Self -Assembly and Deposition Techiniques* (Springer, 2012). If we use a wavelength longer than the absorption edge wavelength, higher efficiency of the photochemical reactions can be expected in various applications, including water splitting and artificial photosynthesis, using sunlight, because most of the materials used in such applications are limited in the ultraviolet regions.

The reasons why longer wavelengths could be used at the nanoscale were investigated and revealed in collaboration with the late Prof. Katsuyuki Nobusada at Institute for Molecular Science. With the help of his deep physical considerations and computational studies based on a first-principles computational program including the ONF effect, one of the key features of photochemical reactions at the nanoscale was identified to be the "non-uniformity" of the optical field. This book focuses on these findings.

The author wishes to express special thanks to the late Prof. Katsuyuki Nobusada at Institute for Molecular Science. However, the author could not continue his research in this direction. He also thanks students in his group Drs. Maiku Yamaguchi, Felix Brandenburg (The University of Tokyo); Ms. Etsuko Ota, Ms. Ayako Mizushima (The University of Tokyo); Prof. Kenji Iida (Hokkaido University); Prof. Kazuhiro Yabana, Dr. Masashi Noda (University of Tsukuba); Profs. Masaki Sekino, Shigeo Maruyama, Shohei Chiashi, Moriaki Kusakabe, Dr. Akihiro Kuwahata

(The University of Tokyo); Prof. Olivier Soppera, Dr. Benjamin Leuschel (Institut de Science des Materiaux de Mulhouse), Prof. Gyu-Chul Yi (Seoul National University), Prof. Fedor Jelezko (Ulm University), Prof. Motoichi Ohtsu (The University of Tokyo) and his collaborators; Prof. Naoya Tate (Kyusyu University); Profs. Mutsuko Hatano, Takayuki Iwasaki, Mamoru Yoshimoto (Tokyo Institute of Technology), Drs. Ryuji Igarashi, Takeshi Ohshima, Yuta Masuyama (National Institutes for Quantum and Radiological Science and Technology), Drs. Satoshi Yamasaki, Daisuke Takeuchi, Toshi-haru Makino, Masahiko Ogura, Hiromitsu Kato, Hideyo Okushi (National Institute of Advanced Industrial Science and Technology), Dr. Satoshi Koizumi (National Institute for Materials Science), Prof. Tatsuki Morimoto (Tokyo University of Technology); Prof. Tomoko Yoshida, Dr. Muneaki Yamamoto (Osaka City University); and Prof. Yuichi Negishi (Tokyo University of Science).

This work was partially supported by the MEXT Quantum Leap Flagship Program (MEXT Q-LEAP) Grant Number JPMXS0118067395, JSPS KAKENHI (Nos. JP23686052, JP24656224, JP26286022, JP26630122, JP15H00866, JP17H01262, JP18H01470, and JP18H05157), the MEXT as a social and scientific priority issue (Creation of new functional devices and high-performance materials to support next-generation industries GCDMSI) to be tackled by using the post-K computer (ID: hp170250), the MEXT Nanotechnology Platform (No. 12024046), Japan (JSPS)-Korea (NRF) Bilateral Program, Japan (JSPS)-France (MAEDI) Bilateral Program SAKURA, the JSPS Core-to-Core Program (A. Advanced Research Networks), Asahi Glass Foundation, Research Foundation for Opto-Science and Technology. Finally, Mr. Ravivarman Selvaraj of Springer Nature is acknowledged for his guidance and suggestions.

Tokyo, Japan Takashi Yatsui

Contents

About the Author

Takashi Yatsui received his B.E. degree from Keio University, Tokyo, Japan, in 1995 and his M.E. and D.E. degrees from the Tokyo Institute of Technology in 1997 and 2000, respectively. From 1999 to 2000, he was a research fellow of the Japan Society for the Promotion of Science and, from 2000 to 2008, a researcher at the Japan Science and Technology Agency, Tokyo. In 2008, he joined the University of Tokyo as an associate professor. In 2020, he joined Toyohashi University of Technology as a professor. His current research interests include nanofabrication using optical near fields and its application to nanophotonics.

He is the recipient of numerous awards for his outstanding scientific achievements. These include first prize in a Paper Contest of the IEEE Student Branch at the Tokyo Institute of Technology in 1998; an Excellent Research Presentation Award from the Japan Society of Applied Physics in 2000; the Tejima Doctoral Dissertation Award from the Tejima Foundation in 2001; the Gottfried Wagener Prize in 2010; the Osaka University Kondo Prize in 2012; the Erlangen Innovation Award for Optical Technologies in 2012; the Fujisankei Business i Award in 2013; the 48th Ichimura Prize in Science from The New Technology Development Foundation in 2016; the APEX/JJAP Editorial Contribution Award in 2016; and the Optical Engineering Award (Takawano Award) from JSAP in 2019.

Chapter 1
Introduction

The demand for a higher areal density in hard disk (HDD) has been rising every day. The required areal density is expected to reach as much as 10 Tb/in^2 [1], in which unit bit size becomes less than 10 nm in diameter. This size is below the wavelength of an extreme ultraviolet (EUV) light source whose wavelength is 13.5 nm. In order to realize a higher spatial resolution over the diffraction limit of light, we need the optical near-field (ONF).

The concept of the ONF, instead of propagating far-field (FF) light, was proposed about a century ago [2]. It is expected to realize super-resolution beyond the diffraction limit by using the localized electric fields generated in the nanoscale structure, such as a metallic aperture. The components of the ONF is expressed as the dipole radiation field from the dipole moment [3] (see the details in Chap. 2). Recent developments in plasmonics [4] and metamaterial [5] often utilize resonances of metalic nanoparticles or periodic structures. Thus, optical loss is indispensable.

In addition to the plasmon resonance, the ONF has unique phenomena even at non-resonant conditions, including dipole forbidden transitions wave vector excitation. The author has been interested in these unique phenomena of the ONF and introduced in the book, $Nanophotonic\ Fabrication : Self - Assembly\ and\ Deposition\ Techniques$ (Springer, 2012) [6]. This book reviews the recent progress of the theoretical studies based on a first-principles computational program including the ONF effect, which revealed that such unique phenomena are originated from the non-uniformity of the ONF (Chap. 2). Based on the theoretical studies, this book reviews the recent development of nanophotonic fabrication techniques for the realization of ultra-flat surface (Chap. 3), new chemical reactions and excitations (Chap. 4), and high-performance photo-detectors (Chap. 5).

T. Yatsui, *Nanophotonic Chemical Reactions*, Nano-Optics and Nanophotonics,
https://doi.org/10.1007/978-3-030-42843-3_1

References

1. http://idema.org/?page_id=5868
2. E.H. Synge, A suggested methode for extending microscopic resolution into the ultra-microscopic region. Philos. Mag. **6**, 356–362 (1928)
3. J.D. Jackson, *Classical Electrodynamics* (Wiley, Hoboken, 1962)
4. S.A. Maier, *Plasmonics: Fundamentals and Applications* (Springer, Berlin, 2007)
5. R.W. Ziolkowski, N. Engheta, *Metamaterials: Physics And Engineering Explorations* (Wiley, Hoboken, 2006)
6. T. Yatsui, *Nanophotonic Fabrication: Self-Assembly and Deposition Techniques* (Springer, Berlin, 2012)

Chapter 2
Non-uniform Optical Near-Field

2.1 Introduction

The optical near-field (ONF) is an electric field, which is localized at nanoscale materials. The ONF is known as local field components from the dipole radiation field. The dipole radiation field from the dipole moment $\boldsymbol{p}$ with an angular frequency ω (=ck ; c is the speed of light and k is the wave number) is given by [1]

$$\boldsymbol{E} = \frac{1}{4\pi\varepsilon_0} \times \left\{ k^2 (\boldsymbol{n} \times \boldsymbol{p}) \times \boldsymbol{n} \left(\frac{1}{r} \right) + [3\boldsymbol{n} (\boldsymbol{n} \cdot \boldsymbol{p}) - \boldsymbol{p}] \left(-\frac{ik}{r^2} + \frac{1}{r^3} \right) \right\} e^{ikr} \tag{2.1}$$

where $\boldsymbol{n}$ is the unit vector along $\boldsymbol{r}$, ε_0 is the permittivity of vacuum, and μ_0 is the permeability of vacuum. In these equations, the far-field components are given by

$$\boldsymbol{E} = \frac{\omega}{k} \boldsymbol{B} \times \boldsymbol{n} \tag{2.2}$$

$$\boldsymbol{B} = \frac{\omega \mu_0 k}{4\pi} (\boldsymbol{n} \times \boldsymbol{p}) \frac{e^{ikr}}{r} \tag{2.3}$$

and the near-field components are given by

$$\boldsymbol{E} = \frac{1}{4\pi\varepsilon_0} [3\boldsymbol{n} (\boldsymbol{n} \cdot \boldsymbol{p}) - \boldsymbol{p}] \frac{1}{r^3} \tag{2.4}$$

$$\boldsymbol{B} = \frac{i\omega \mu_0}{4\pi} (\boldsymbol{n} \times \boldsymbol{p}) \frac{1}{r^2} \tag{2.5}$$

T. Yatsui, *Nanophotonic Chemical Reactions*, Nano-Optics and Nanophotonics,
https://doi.org/10.1007/978-3-030-42843-3_2

As shown in (2.4), the near-field components of the electric field are proportional to r^{-3}, indicating that the ONF has a spatial "non-uniformity" [2]. In this section, we review some unique effects originated from this non-uniformity.

2.2 Photodissociation via Vibrational Levels

We examined how the non-uniformity of the ONF results in photodissociation processes. As of photodissociation processes, we examined of the molecular dissociations, which are important in photochemical reactions [3]. Most of the experiment results of photodissociation are understood using uniform electric field from the propagating far-field light. As well as, the numerical simulations of the photodissociation processes use the dipole approximation, where the electric field has uniform spatial distribution. More recently, many groups used non-uniform ONFs to understand molecular excitation processes, including Kim et al. [4]. As described in [4], high harmonic generation (HHG) has been reported by the generation of the ONF at gold nanostructures with a bow tie shape.

To make it clear of the effect of the non-uniformity in ONFs, we investigated the dissociation process of $H_2{}^+$ via a non-uniform ONF [5]. $H_2{}^+$ is one of the good molecules for experimental and theoretical studies on photodissociation [6–10].

As is described in [11], we used a reduced-dimensional model to represent $H_2{}^+$. To describe the system, we used two coordinates R and z, which are the distance between the nuclei and the distance from the midpoint of the nuclei to the electron, respectively. The static Hamiltonian of the model, $\hat{H}_0$, is written as

$$\begin{aligned}\hat{H}_0(z, R) = &-\frac{1}{2\mu_z}\frac{\partial^2}{\partial z^2} - \frac{1}{2\mu_R}\frac{\partial^2}{\partial R^2} - \frac{1}{\sqrt{(z-R/2)^2+\alpha}} \\ &- \frac{1}{\sqrt{(z+R/2)^2+\alpha}} + \frac{1}{\sqrt{R^2+\beta}},\end{aligned} \tag{2.6}$$

where $\mu_R = m_p/2$ and $\mu_z = 2m_p m_e/(2m_p + m_e)$ are reduced masses calculated from the proton mass m_p and the electron mass m_e. Screening parameters α and β introduced to avoid the divergence of potential were set to be 1.0 and 0.03, respectively.

In order to investigate the time-dependent photodissociation process of $H_2{}^+$, we solved the time-dependent Schrödinger equation in atomic units

$$i\frac{\partial}{\partial t}\Psi(z, R, t) = \left[\hat{H}_0(z, R) + \hat{V}_{\text{ext}}(z, R, t)\right]\Psi(z, R, t), \tag{2.7}$$

where $\hat{V}_{\text{ext}}$ represents light excitation. Here, we used the simple model of the electric field, in which the amplitude of a non-uniform ONF has a linear function

$$E(z) = E_0\,(1 + 2\epsilon z)\,, \tag{2.8}$$

where E_0 is the field amplitude at the midpoint of the nuclei and ϵ is defined as $E'(z = 0)/2E(0)$ and represents the extent of non-uniformity. As described in Sect. 2.1, the ONF was approximated as electric dipole radiation. However, here we used (2.8) to represent a non-uniform ONF. This is because that the electric field generated from an electric dipole diverges at the dipole position. Therefore, we need to place the dipole outside of the simulation box. However, if we place the dipole apart from the simulation box, it results in the large distance between the dipole and H_2^+ to evaluate the effect of a non-uniform field when a sufficiently large simulation box is employed. Another reason is that if we use the (2.8), we could investigate the effect of field non-uniformity directly by using various values of ϵ in the calculations. By integrating (2.8) and multiplying the time-dependent term, $\hat{V}_{\text{ext}}$ is obtained as

$$\begin{aligned}\hat{V}_{\text{ext}}(z, R, t) &= \left[\int_0^z dz' E(z') - \int_0^R dz' \left\{E\left(\frac{z'}{2}\right) - E\left(-\frac{z'}{2}\right)\right\}\right] \\ &\quad \times \sin(\omega t) f(t) &(2.9)\\ &= E_0\left[z(1+\epsilon z) - \epsilon R^2\right] \sin(\omega t) f(t), &(2.10)\end{aligned}$$

where ω is the photon energy of the electric field, and $f(t)$ is an envelope function of the incident light. As for $f(t)$, we used a trapezoidal envelope function

$$f(t) = \begin{cases} \sin^2\left(\frac{\pi t}{2T_1}\right) & (0 \le t < T_1) \\ 1 & (T_1 \le t < T_2) \\ \sin^2\left(\frac{\pi(t-(2T_2-T_3))}{2(T_3-T_2)}\right) & (T_2 \le t < T_3) \\ 0 & (T_3 \le t) \end{cases}, \tag{2.11}$$

where we set the parameters as $T_1 = 10$ fs, $T_2 = 30$ fs, and $T_3 = 40$ fs. The first and second integrals in (2.9) indicate the respective potentials exerted on the electron and nuclei. If the field has non-uniform distribution ($\epsilon \neq 0$), $\hat{V}_{\text{ext}}$ depends on the R.

We solved the time-dependent Schrödinger equation with an improved Crank–Nicolson scheme [12] in which a time spacing was set to $dt = 0.0005$ fs. The simulation area was $|z| \le 80$ a.u. and $0 \le R \le 15$ a.u. with spacings of $dz = 0.2$ a.u. and $dR = 0.05$ a.u. We used a mask function of $\cos^{\frac{1}{6}}$ at the boundaries of both the z and R coordinates, so that the emitted wave packets were absorbed. The wave packets absorbed by the mask function at the z- and R-boundaries were considered as ionized and dissociated, respectively.

Figure 2.1a shows the Born–Oppenheimer potential energy surface, in which H_2^+ has a bound ground state ($1\sigma_g$) and a repulsive excited state ($2\sigma_u$). There are 20 vibrational states for the $1\sigma_g$ state. We set the vibrational ground state $|v = 0\rangle$ as for the initial state in the calculation.

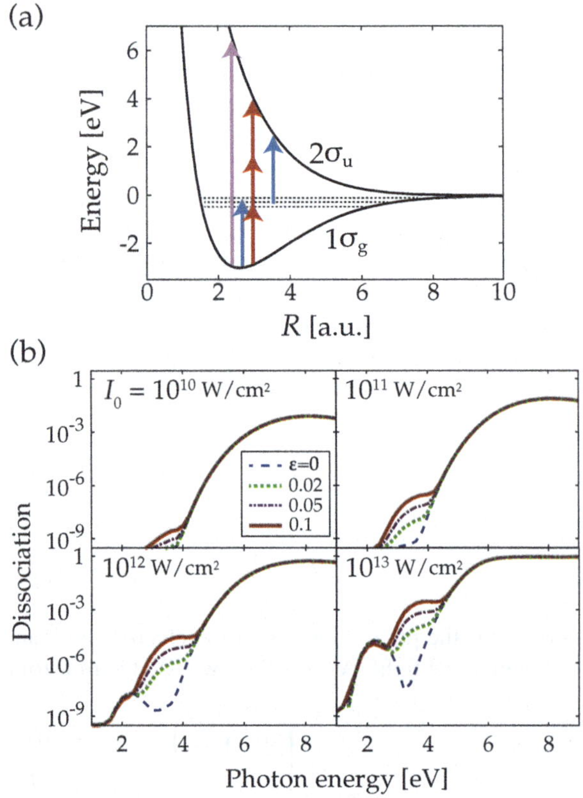

Fig. 2.1 **a** Born–Oppenheimer potential energy surface of $H_2{}^+$. The red, blue, and purple arrows represent three dissociation paths. **b** Dissociation spectra of $H_2{}^+$ for light intensities $I_0 = \{10^{10}, 10^{11}, 10^{12}, 10^{13}\}$ W/cm^2 and field non-uniformities $\epsilon = \{0, 0.02, 0.05, 0.1\}$. Adapted from [5] Copyright 2016, APS Journals

We calculated the dissociation spectra (see Fig. 2.1b) for various values of I_0 (incident light intensities) and ϵ (non-uniformities). Note that we obtained three peaks at around 2.0, 3.6, and 8.0 eV for $I_0 > 10^{12}$ W/cm^2. Because the 8.0 eV corresponds to the energy difference between excited state and ground state (indicated as the purple arrow in Fig. 2.1a), this energy peak is originated from the one-photon absorption. Because the one-photon absorption is a linear optical process, we could observe the peak at 8.0 eV in the lowest incident power at $I_0 = 10^{12}$W/cm^2. The peak at 2.0 eV is considered to be originated from a three-photon process (indicated as red arrows in Fig. 2.1a). This process has been investigated experimentally and numerically [6–10, 13]. By comparing the absorption spectra depending on the ϵ values, these two dissociation paths at 2.0 and 8.0 eV remained the same as the ϵ value increased, meaning that those two dissociation paths have no dependence on spatial non-uniformity of the electric field. In contrast, we observed that the peak at 3.6 eV strongly depends on the ϵ values, in which there is no peak at $\epsilon = 0$, i.e., uniform electric field, while the peak value increased rapidly as ϵ value increased. Through the below discussions, we concluded that the peak at 3.6 eV originates from the two-step transition via vibrational excited states (indicated as the blue arrows in Fig. 2.1a).

As is described in many literatures [14–18], it is known that the non-uniform field induces electric quadrupole transitions, which are forbidden under the electric dipole approximation. Similarly, we show that a non-uniform field can induce vibrational excitation processes that are not allowed by uniform field. A transition matrix element between vibrational states, denoted as v and v', is calculated by

$$\left\langle v'\right| \hat{V}_{\mathrm{ext}} \left|v\right\rangle = \iint dzdR\, \phi^{*}_{v'}(z, R)\hat{V}_{\mathrm{ext}}(z, R)\phi_v(z, R), \tag{2.12}$$

where ϕ_v is the wavefunction of the vibrational state $|v\rangle$. By substituting $\hat{V}_{\mathrm{ext}}$ given by (2.10) into (2.12) except for the time-dependent part, the matrix element is obtained as the sum of three parts:

$$\begin{aligned}\left\langle v'\right| \hat{V}_{\mathrm{ext}} \left|v\right\rangle = E_0 &\iint dzdR\, \phi^{*}_{v'}(z, R)z\phi_v(z, R)\\ + \epsilon E_0 &\iint dzdR\, \phi^{*}_{v'}(z, R)z^2\phi_v(z, R)\\ - \epsilon E_0 &\iint dzdR\, \phi^{*}_{v'}(z, R)R^2\phi_v(z, R).\end{aligned} \tag{2.13}$$

Note that the second and third terms are proportional to ϵ. Since ϕ_v for all v belongs to the $1\sigma_g$ state, it is an even function of z. This makes the integrand of the first term of (2.13) an even function; thus, the value of the first integral becomes zero. In contrast, the second and third terms are, in general, nonzero. Therefore, it follows that the transition matrix element $\left\langle v'\right| \hat{V}_{\mathrm{ext}} \left|v\right\rangle$ is proportional to ϵ. Thus, the transition rate from $|v\rangle$ to $\left|v'\right\rangle$, $T_{v\to v'}$, depends on ϵ as

$$T_{v\to v'} \propto \left|\left\langle v'\right| \hat{V}_{\mathrm{ext}} \left|v\right\rangle\right|^2 \propto \epsilon^2. \tag{2.14}$$

This dependence indicates that a non-uniform ONF with a finite ϵ causes transitions from the vibrational ground state to excited states while uniform light with $\epsilon = 0$ does not cause this transition, as is generally known.

To clarify the dissociation process, we obtained the dependence on the intensity and non-uniformity of the incident light for three-photon energies discussed above (2.0, 3.6, and 8.0 eV). Figure 2.2a shows the dependence on light intensity I_0 at $\epsilon = 0.1$. From the results, the power dependences of the dissociation at 8.0, 3.6, and 2.0 eV are found to be proportional to I_0, ${I_0}^2$, and ${I_0}^3$, respectively. These dependences confirmed that they corresponded to one-, two-, and three-photon processes, respectively (Fig. 2.1a). Figure 2.2b shows the ϵ (the field non-uniformity) at $I_0 = 10^{12}\ \mathrm{W/cm^2}$. From this result, it was found that the dissociation for 2.0 and 8.0 eV has constant as a function of ϵ. In contrast, that for 3.6 eV was proportional

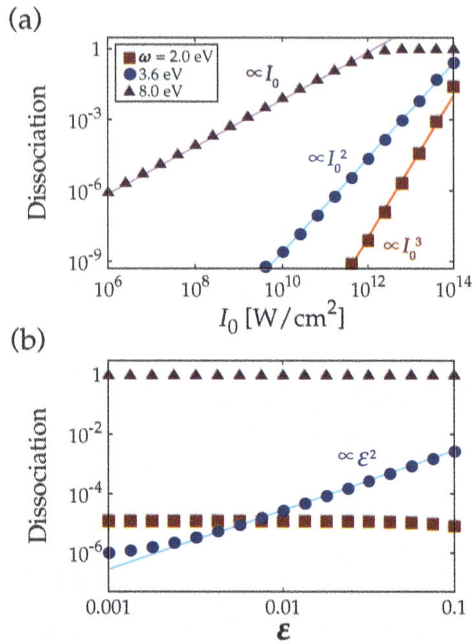

Fig. 2.2 Dependence of dissociation caused by non-uniform light with photon energies of 2.0 eV (red squares), 3.6 eV (blue circles), and 8.0 eV (purple triangles) on **a** light intensity I_0 and **b** field non-uniformity ϵ. Adapted from [5] Copyright 2016, APS Journals

to ϵ^2, which is consistent with (2.14). The dissociation process originated from the non-uniform field consists of two transitions:

(i) from the vibrational ground state to vibrationally excited states
(ii) from vibrationally excited states to the $2\sigma_u$ state

Transition (ii) is electric dipole allowed, therefore the rate should be constant as a function of ϵ. Therefore, the dependence of the dissociation rate on ϵ at 3.6 eV is determined by transition (i) and hence is proportional to ϵ^2 according to (2.14).

2.3 Second Harmonic Generation by the Non-uniform ONF

We examined how the non-uniformity of the ONF results in generation of the HHG [19, 20]. There is a selection rule that even order of HHG is not excited in materials with inversion symmetry by the propagating far-field light, in which the field distribution is uniform. This is explained in the following. When the material is excited by the electric field $E(t)$ with the uniform field distribution, the dielectric polarization $P(t)$ is generated as [21]

$$P\,(t) = \alpha E\,(t) + \beta E(t)^2 + \gamma E(t)^3 + \delta E(t)^4 + \cdots \tag{2.15}$$

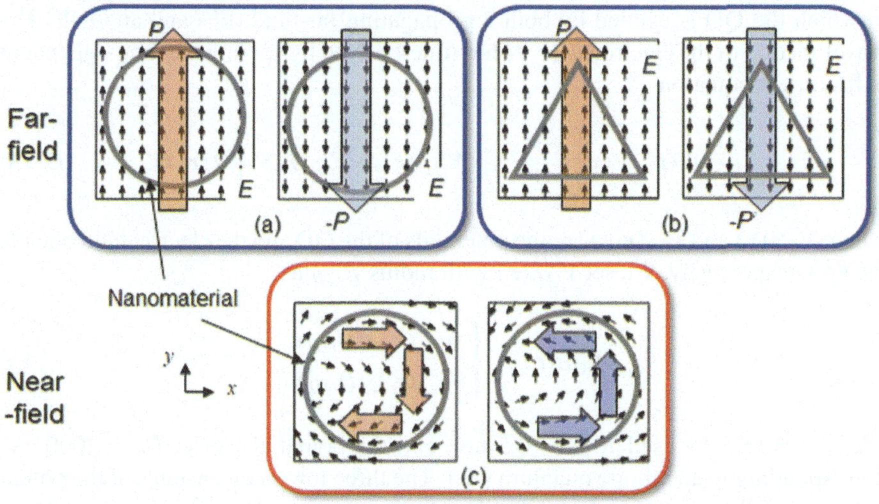

Fig. 2.3 Snap shot of the excited polarization, P. **a** Far-field excitation in the material with inversion symmetry. Gray solid circle represents the outline of the nanomaterial. **b** Far-field excitation in the material without inversion symmetry. Gray solid triangle represents the outline of the nanomaterial. **c** Near-field excitation in the material with inversion symmetry

Now we consider that the nanomaterial with inversion symmetry is excited by the electric field $E(t)$ (see Fig. 2.3a). When the $E(t)$ is applied along y, the excited polarization is $P(t)$ as shown in (2.15). If the $E(t)$ is applied along $-y$, the excited polarization is expressed as follows

$$-P(t) = -\alpha E(t) + \beta E(t)^2 - \gamma E(t)^3 + \delta E(t)^4 - \cdots \tag{2.16}$$

Then if we add (2.15) and (2.16), we obtain

$$0 = \beta E(t)^2 + \delta E(t)^4 + \cdots \tag{2.17}$$

This (2.17) indicates that the even orders of HHG including second harmonic generation (SHG) should be zero.

Thus, the generation of SH light requires nanomaterials without inversion symmetry (Fig. 2.3b). However, if the materials have inversion asymmetry (Fig. 2.3b), the excited polarization $|P'|$ is not equal to $|P|$. Therefore, the generation of even orders of HHG is allowed.

In contrast, if the nanomaterial is excited by the non-uniform ONF (Fig. 2.3c), the excited polarization is not uniform even in the nanomaterial with inversion symmetry. Then, the generation of even orders of HHG is allowed as same as that in the materials with inversion asymmetry.

To investigate more quantitatively, we performed numerical calculations. In the calculation, we used an electron in a two-dimensional quantum dot (QD) model, in which the QD is excited by both a propagating far-field (FF) and an ONF. The wavefunction of the electron $\Psi(\boldsymbol{r}, t)$ has time evolved by solving the time-dependent Schrödinger equation

$$i\frac{\partial}{\partial t}\Psi(\boldsymbol{r}, t) = \left[-\frac{1}{2}\nabla^2 + V_{QD}(\boldsymbol{r}) + V_{\text{ext}}(\boldsymbol{r}, t)\right]\Psi(\boldsymbol{r}, t), \tag{2.18}$$

where $V_{QD}(\boldsymbol{r})$ and $V_{\text{ext}}(\boldsymbol{r}, t)$ are the potentials of the QD and the electric field outside the QD, respectively. We set $V_{QD}(\boldsymbol{r})$ with radius R_{QD} as

$$V_{QD}(\boldsymbol{r}) = \begin{cases} 0 & (r < R_{QD}) \\ U_0 & (r \geq R_{QD}) \end{cases}, \tag{2.19}$$

where $r = |\boldsymbol{r}|$. We used $R_{QD} = 0.5$ nm, and the potential barrier $U_0 = 2000$ eV, corresponding to an infinite quantum well. The three lowest eigenstates of the potential $V_{QD}(\boldsymbol{r})$ are shown in Fig. 2.4a. The energy spacing between the ground and first excited states of $\psi_{ex}(\boldsymbol{r})$ and $\psi_{ey}(\boldsymbol{r})$ (ΔE) is 1.33eV, in which the first excited states ($\psi_{ex}(\boldsymbol{r})$ and $\psi_{ey}(\boldsymbol{r})$) are twofold degenerate. The transition dipole moments $\langle\psi_{ex}|\,\hat{\mu}_x\,|\psi_g\rangle$ and $\langle\psi_{ey}|\,\hat{\mu}_y\,|\psi_g\rangle$ are nonzero, where $\hat{\mu}_x$ and $\hat{\mu}_y$ are the x- and y- components of the electric dipole operator $\hat{\boldsymbol{\mu}}$, respectively.

$V_{\text{ext}}(\boldsymbol{r}, t)$ is the time-dependent external potential. Figure 2.4b, c shows the schematic of FF and NF excitation. The $V_{\text{ext}}(\boldsymbol{r}, t)$ for FF excitation is given by

$$V_{\text{ext,far}}(\boldsymbol{r}, t) = -\boldsymbol{r} \cdot \boldsymbol{E}(t), \tag{2.20}$$

where $\boldsymbol{E}(t)$ is the electric field amplitude of the FF light. $\boldsymbol{E}(t)$ has uniform spatial distribution over the QD. As for ONF excitation, $V_{\text{ext}}(\boldsymbol{r}, t)$ is given by [14, 22, 23]

$$V_{\text{ext,near}}(\boldsymbol{r}, t) = -\frac{\boldsymbol{r} \cdot \boldsymbol{P}(t)}{|\boldsymbol{r} - \boldsymbol{R}_p|^3}, \tag{2.21}$$

where $\boldsymbol{P}(t)$ is a dipole moment, and $\boldsymbol{R}_p$ is the coordinate of the dipole source.

Figure 2.4d, e shows the schematic of $V_{\text{ext}}(\boldsymbol{r})$ of the FF and of the ONF, respectively. As we can see, both $\boldsymbol{E}(t)$ and $\boldsymbol{P}(t)$ are polarized along the y-axis. The white arrows show the opposite direction of the electric field $\boldsymbol{E}$, which is the force exerted on an electron at each point. It is noted that the electric fields of the FF and ONF are uniform and non-uniform, respectively. Furthermore, although the FF has only the y-components, the ONF has the x-component by the y-polarized dipole source (Fig. 2.4e).

$\boldsymbol{E}(t)$ and $\boldsymbol{P}(t)$ have time-dependent function as

$$\boldsymbol{E}(t) = \boldsymbol{E}_0 \sin(\omega t) f(t) \tag{2.22}$$

$$\boldsymbol{P}(t) = \boldsymbol{P}_0 \sin(\omega t) f(t), \tag{2.23}$$

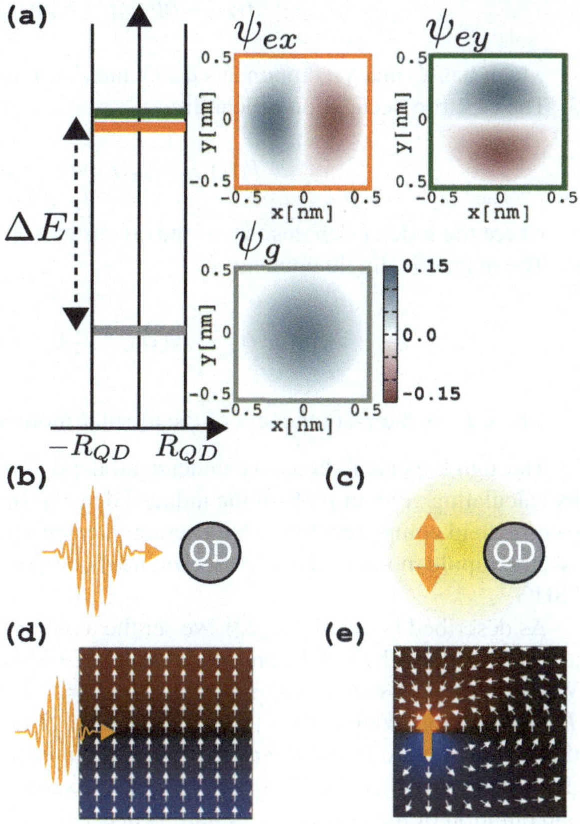

Fig. 2.4 **a** Energy diagram of eigenstates of the two-dimensional QD model. R_{QD}:QD radius, ΔE: the energy spacing between the ground state (ψ_g) and the first excited states ($\psi_e x$ and $\psi_x y$). Right side images show spatial distributions of the real parts of the wavefunctions. Schematic of **b** FF and **c** NF excitation of QD. Schematic of the potentials for **d** the FF excitation and **e** the NF excitation. The red and blue colors in the contour plots represent high and low values of the potential, and the white arrows indicate the direction of the force exerted on the electron at each position. Adapted from [20] Copyright 2015, APS Journals

where $\boldsymbol{E}_0$ and $\boldsymbol{P}_0$ are the amplitudes of the electric field and the electric dipole, respectively, ω is the angular frequency of the external field, and $f(t)$ is the pulse envelope function:

$$f(t) = \exp\left(-\frac{(t-\tau_c)^2}{2{\tau_d}^2}\right), \tag{2.24}$$

where τ_c and τ_d are the center of the pulse and the pulse duration, respectively.

We calculated four values.

1. The induced density of the electron:

$$\Delta\rho(\boldsymbol{r}, t) = \rho(\boldsymbol{r}, t) - \rho(\boldsymbol{r}, 0), \tag{2.25}$$

   where $\rho(\boldsymbol{r}, t)$ is the electron density at point $\boldsymbol{r}$ and time t, calculated by $\rho(\boldsymbol{r}, t) = |\Psi(\boldsymbol{r}, t)|^2$.
2. The induced electric dipole moment:

$$\boldsymbol{d}(t) = \langle \Psi(t)| - \boldsymbol{r} \, |\Psi(t)\rangle \, , \tag{2.26}$$

which has x- and y-components $d_x(t)$ and $d_y(t)$, respectively.

3. The electron occupation probability:

$$P_i(t) = |\langle \psi_i(r) | \Psi(\boldsymbol{r}, t)\rangle|^2, \tag{2.27}$$

where the index i denotes one of the eigenstates g, ex, or ey drawn in Fig. 2.4a.

4. The magnetic dipole moment:

$$\mu_z(t) = \langle \Psi(t)| - \frac{1}{2}\hat{L}_z \, |\Psi(t)\rangle \, , \tag{2.28}$$

where $\hat{L}_z$ is the z-component of the angular momentum operator.

Through the calculations, we understand the dynamics of the electron excitation by calculating contour plots of the induced density. In addition, we visualize the SH electric field component in the NF excitation calculating the time dependence of the electric dipole moment. The SH electric field component of the ONF resulted in the "SHG.'

As described in details in [20], we set the angular frequency of the external field ω to be 0.665 eV (half of the energy spacing ΔE of the QD) to verify the SHG effect. We set the dipole source with a separation $R_p = 0.7$ nm from the QD. Both $\boldsymbol{P}_0$ and $\boldsymbol{E}_0$ have y-polarization with $|\boldsymbol{P}_0| = 5.0$ D and $|\boldsymbol{E}_0| = 4.4 \times 10^9$ V/m. We chose these values to produce the same electric field intensity at the position of the QD so that we can compare the FF and NF different excitations directly. We set the center and duration of the pulse as $\tau_c = 45.0$ fs and $\tau_d = 17.5$ fs, respectively.

We compared the time dependence of electron excited by the FF (left-hand side) and the NF (right-hand side) excitations (see Fig. 2.5). Figure 2.5a shows the time-dependent spatial distributions of the induced electron density at various time t of $\Delta\rho(\boldsymbol{r}, t)$ ($t = 19.2, 31.6, 44.1, 56.5, 68.9$, and 81.4 fs) with the phase of the applied electric fields becomes $2\pi N + \pi/6$ (N is integer). As can be seen in Fig. 2.5a, the electron density by the FF excitation has an oscillation only along the y-axis, and the intensity decreases at $t = 81.4$ fs. In contrast, the electron density by the NF excitation has rotational oscillation of electron density from y-axis to x-axis, and the excitation remains after the pulse excitation. The oscillation along the x-axis reflects the fact that the NF generated by the electric dipole moment with the y-polarization has the x-component (see Fig. 2.4e).

The black solid lines in the upper panels of Fig. 2.5b are the external (FF and NF) electric fields to excite the QD. Because the NF component of the electric field intensity depends on the distance from the source, we need to define the position where the ONF intensity is specified and then compare the time dependence of electron induced by the same intensity for both FF and NF excitation on an equal condition. Regarding NF intensity, we select the field amplitude at the position of

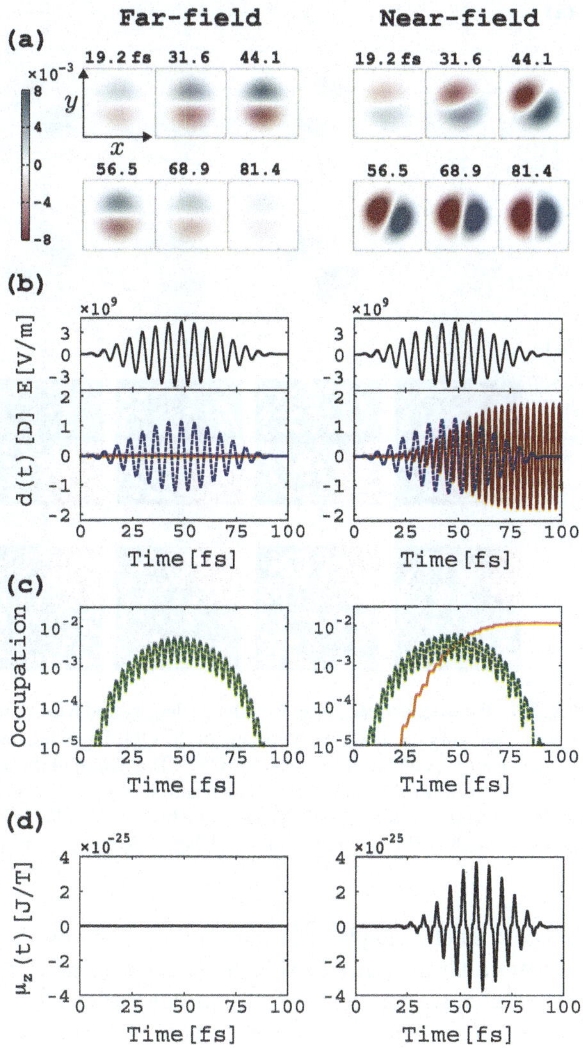

Fig. 2.5 Time dependence of electron excited by the FF (left-hand side) and NF (right-hand side). **a** The spatial distributions of the induced density, $\Delta\rho(\boldsymbol{r}, t)$, at $t = 19.2, 31.6, 44.1, 56.5, 68.9$, and 81.4 fs. **b** The excited electric field (upper) and the induced electric dipole moments (lower). The red solid line: x-component of the dipole moments, blue-dashed lines indicate: the y-component of the dipole moments. **c** The electron occupation probabilities of the excited states. The orange solid (ψ_{ex}) and green-dashed (ψ_{ey}) lines indicate the occupations. **d** Induced z-components of the magnetic dipole moment μ_z. Adapted from [20] Copyright 2015, APS Journals

the QD center and set the intensity same value of the FF. We found that the both external fields of FF and NF excitations result in equivalent electric field intensities.

The lower panels of Fig. 2.5b show the x- and y-components of the dipole moments $d_x(t)$ (the red solid line) and $d_y(t)$ (the blue-dashed line) with FF and NF excitations, respectively. The y-components of dipole moments $d_y(t)$ induced by both the FF and NF were almost the same of the incident electric fields, and they decreased as the incident pulse decreased. This time dependence is originated from the non-resonance of the incident frequency (=0.665 eV) with the resonant value of the energy spacing between the ground state and the first excited state (=1.33 eV. See Fig. 2.4a). However, x-components of dipole moments $d_x(t)$ do not synchronize the

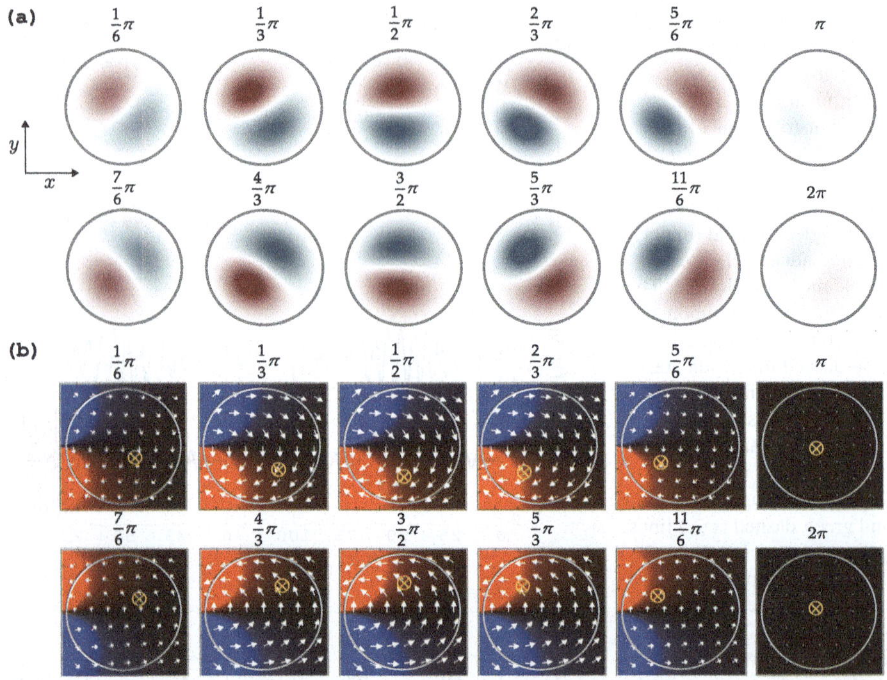

Fig. 2.6 Time dependence of **a** the induced density and **b** the potential induced by the electric dipole source. The gray circle shows the shape of the QD. The white arrows represent the direction of the force that applies on the electron by the ONF. The length of the arrows corresponds to the amplitude of the dipole source at each phase. The position of yellow ⊗ corresponds to the expectation value of the electron position $\langle \Psi(t)| \boldsymbol{r} |\Psi(t)\rangle$, in which the displacement from the center of the QD is ten times as much as the calculated value. Adapted from [20] Copyright 2015, APS Journals

incident electric fields: The intensity of $d_x(t)$ by the NF excitation increases and remains after the pulse ends, and the intensity of $d_x(t)$ by the FF excitation is zero. In addition, we found that the frequency of the $d_x(t)$ induced by the NF is twice as much as that of the incident electric field ω. That is, the frequency of $d_x(t)$ oscillation is 2ω $(= \Delta E)$. The fact that the dipole moment has second harmonic oscillation only by the NF excitation supports that the SHG is originated by the non-uniform ONF.

To validate the SHG, we plotted the time dependence of the electron occupation probabilities in Fig. 2.5c. The orange solid and green-dashed lines represent $P_{ex}(t)$ and $P_{ey}(t)$, the occupations of ψ_{ex} and ψ_{ey}, respectively, in which the respective colors are the same as that of wavefunctions shown in Fig. 2.4a. The occupations of $P_{ey}(t)$ for both of the NF and FF excitations initially enhance in proportion to the incident electric field and decrease as the pulse end. In contrast, the occupations of $P_{ex}(t)$ for the FF and NF excitations have different behaviors. In the FF excitation, $P_{ex}(t)$ is not excited as shown in Fig. 2.5c. This is originated from no x-component of the electric field in the light with y-polarization. However, in the NF excitation, $P_{ex}(t)$ increases and continues even if the pulse ends, in which the NF source has

y-polarization and its angular frequency is not resonant with ΔE. This supports the evidence of the SHG.

SHG by metals including gold and silver is well known owing to electric field enhancement. However, we emphasis that the conventional SHG by metals requires the plasmon resonance at certain wavelength [24, 25]. The scattering and absorption cross sections are explained by point dipole model [26] as follows:

$$C_{\mathrm{sca}} = \frac{k^4}{6\pi}|\alpha|^2 = \frac{8\pi}{3}\left(\frac{2\pi}{\lambda}\right)^4 a^6 \left|\frac{\varepsilon - \varepsilon_m}{\varepsilon + 2\varepsilon_m}\right|^2 \tag{2.29}$$

$$C_{\mathrm{abs}} = k\mathrm{Im}\,[\alpha] = 4\pi\left(\frac{2\pi}{\lambda}\right) a^3 \mathrm{Im}\left[\frac{\varepsilon - \varepsilon_m}{\varepsilon + 2\varepsilon_m}\right] \tag{2.30}$$

where α is the polarizability, a is the particle radius, ϵ and ϵ_{m} are the dielectric function of the particle and the surrounding media, and λ is the wavelength. It is easy to understand from (2.29) that ϵ should have negative value($-2\epsilon_{\mathrm{m}}$) for the resonance, indicating the scattering material needs to be a metal for the plasmon resonance. In addition, (2.30) indicates that the plasmon resonance inherently accompanies a strong optical absorption. However, the SHG by ONFs described above can be obtained by the non-uniformity of the electric field only.

To find the effect of SHG more clearly, we obtained the time dependence of electron excitation during 43.5 fs $\leq t \leq$ 49.7 fs when the source of ONF oscillates. Figure 2.6a shows the spatial distribution of the induced density. In contrast to Fig. 2.5a showing the averaged induced density during several periods, Fig. 2.6a shows the density during a single period by changing phase ϕ. These results reveal that the density rotates in a clockwise direction ($\phi = \pi/6$ to π) and in an anticlockwise direction ($7\pi/6$ to 2π). From these results, the rotational motion repeats every period and gradually becomes strong as time evolute. In addition to the characteristic of rotational motion, as can be seen in Fig. 2.6a, the electron dynamics has another characteristic in the case of NF excitation, i.e., the frequency of the oscillation along the x-axis has twice as much as that of the incident ONF. Indeed, the region of high electron density drawn by the blue region in Fig. 2.6a oscillates twice along the x-axis during one period of the ONF oscillation.

Figure 2.6b shows the contour plots of the NF potential. The respective colors of red and blue correspond to high and low values of the potential, and the white arrows denote the direction (or flow) of $-\boldsymbol{E}$, corresponding to the force exerted on the electron in the same manner as in Fig. 2.4d, e. Now, we mention about the relationship between the rotational motion and the SH component originated from the ONF excitation.

The yellow $\otimes$ in Fig. 2.6b corresponds to the position of the electron $\langle\Psi(t)|\,\boldsymbol{r}\,|\Psi(t)\rangle$ (the displacement from the center of the QD is shown by tenfolds). First, from $\phi = \pi/6$ to $5\pi/6$, the electron locates in the $-y$ region and is subjected to force in the $-x$ direction, and this is found from fact that the arrows around the electron position are left-oriented. After $\phi = \pi$ when the force becomes 0, the direction of the

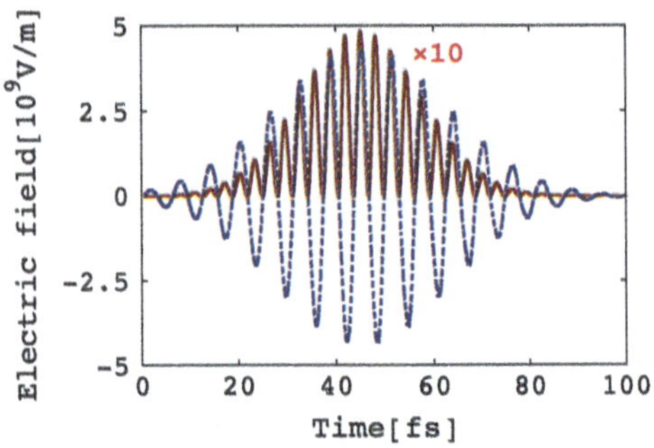

Fig. 2.7 Electric field of the ONF calculated at the expectation value of the electron position $\langle\Psi(t)|\boldsymbol{r}|\Psi(t)\rangle$. Red solid and blue-dashed lines are y- and x-component. Adapted from [20] Copyright 2015, APS Journals

force changes into $-x$ (from $7\pi/6$ to $11\pi/6$) and then becomes 0 again (at $\phi = 2\pi$). From the results, the frequency of the force exerted on the electron along the x-axis has twice as much of the NF source. These discussions reveal the origin of the SH component of electron and the resultant SHG.

We evaluated the electric field of the NF calculated at the position of the expectation value of the electron $\langle\Psi(t)|\boldsymbol{r}|\Psi(t)\rangle$ (see Fig. 2.7). Note that the second harmonic electric field component is inherent in the NF excitation. As can be seen, the frequency of x-component of the electric field (red solid line) clearly has twice as much as of the y-component (blue-dashed line), indicating the SH electric field component of the NF excitation. We also find that the x-component of the electric field has only positive values. Such the asymmetry is originated from the non-uniformity of the NF intensity along the x-axis.

We emphasize that we can bring the SH electric field component generation only when we consider the electron movement as discussed above. If we just calculate the force exerted on the center of the QD, its x-component is zero all the time. Therefore, when we examine the SHG excited by the ONF, it is not enough to calculate the results of the electromagnetic field and static quantum chemical computations. We need to solve the coupled dynamics taking account of explicit light and matter interactions.

As described in detail in [19], the amplitude of the SH electric field component by the ONF excitation is given by

$$E_x^{(2\omega)} = \Delta x \left(\frac{\partial E_{x,\mathrm{near}}}{\partial x}\right)_{r=0} + \Delta y \left(\frac{\partial E_{x,\mathrm{near}}}{\partial y}\right)_{r=0},$$
$$E_y^{(2\omega)} = \Delta x \left(\frac{\partial E_{y,\mathrm{near}}}{\partial x}\right)_{r=0} + \Delta y \left(\frac{\partial E_{y,\mathrm{near}}}{\partial y}\right)_{r=0}, \tag{2.31}$$

where Δx and Δy are the x- and y-components of the displacement from the equilibrium position of the electron, $\boldsymbol{r} = 0$, which correspond to $\langle\Psi(t)|\hat{x}|\Psi(t)\rangle$ and $\langle\Psi(t)|\hat{y}|\Psi(t)\rangle$, respectively, in this study. $E_{x,\mathrm{near}}$ and $E_{y,\mathrm{near}}$ correspond to the electric field amplitudes of the NF, which are calculated from derivatives of the potential in (2.21). Although the y-component of the electric field shown in Fig. 2.7 does not have the SH electric field component because $\nabla E_{y,\mathrm{near}}|_{r=0}$ is completely zero in this arrangement of the system, change in the angle between the electric dipole moment of the dipole source and the line connecting the dipole and the QD can make the y-component of the electric field to also have the SH component. The amplitude of the SH electric field component strongly varies with the configuration of the system in general.

As can be seen in (2.31), the SH component inevitably generates if the incident electric field amplitude has spatial non-uniformity. Therefore, the SHG should be induced as long as ONFs with a non-uniformity, regardless of the ONF source.

2.4 Wave Vector Excitation by the Non-uniform ONF

It is expected that ONF effect could improve the carrier excitation especially in indirect band gap materials. This is because that ONF has large wave vector components (Δk), which implies very small Δx, due to the field localization (Fig. 2.8). This effect was confirmed in a study which ONFs, as a dipole mode plasmon, generate carriers directly [27]. By introducing small particles, the plasmons induce an electric field and resultant Fourier components with large wave numbers. In addition, it was demonstrated that the localized electric field generates carriers without phonon assistance. Furthermore, Jung et al. reported the ONF excitation system in a system comprising metallic nanospheres in Si [28]. In that system, the excitation was evaluated using

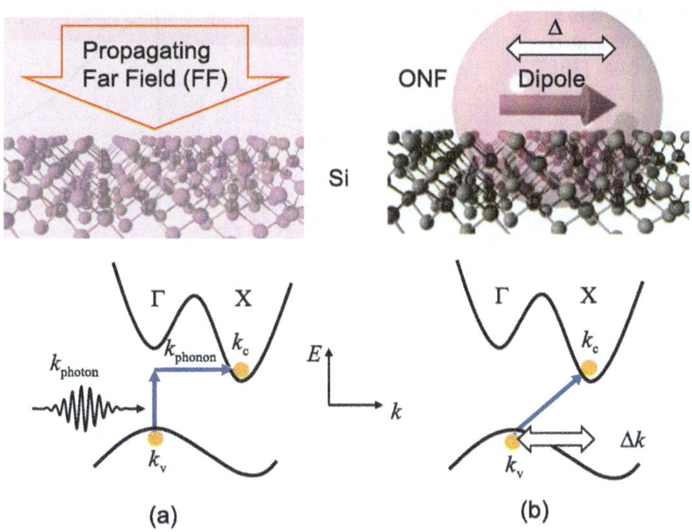

Fig. 2.8 **a** Far-field (FF) excitation of Si. Here k_{photon}, k_c, k_v, and k_{phonon} are the wave vectors of photon, conduction band, valence band, and phonon, respectively. **b** Optical near-field (ONF) excitation of Si. These figures are reprinted from [29] under the terms of the Creative Commons Attribution license

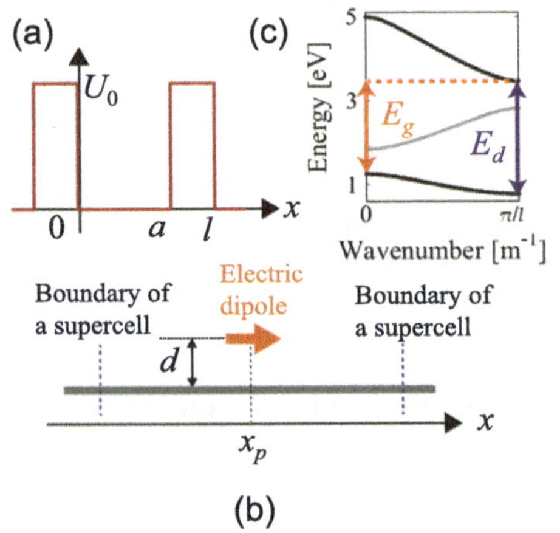

Fig. 2.9 **a** Potential, **b** a schematic of the 1D-KP model, and **c** dispersion relation. Here, E_g: band gap energy and E_d: direct band gap energy. These figures are reprinted with permission from [30], Copyright ©2016 American Physical Society

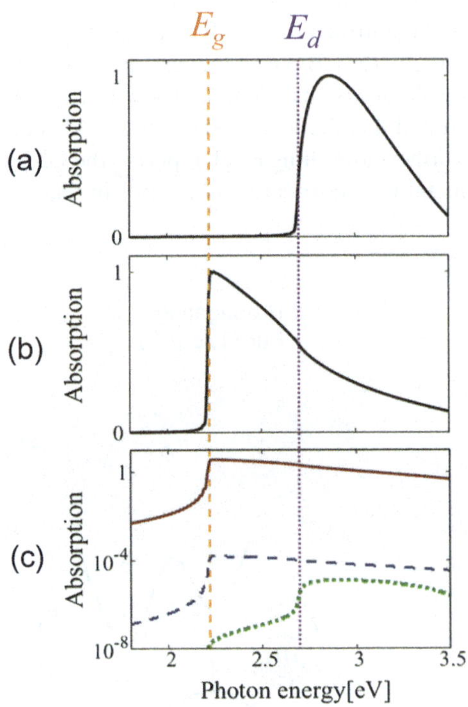

Fig. 2.10 **a** and **b** show the normalized absorption spectra of the 1D-KP model due to the far- and near-field excitation, respectively. **c** Absorption spectra due to the electric field components from the dipole radiation, which are r^{-3} (red solid line), r^{-2} (blue-dashed line), and r^{-1} (green-dotted line). These figures are reprinted with permission from [30], Copyright ©2016 American Physical Society

linear perturbation theory. And then it was found that the ONF effect is too weak to be measured in a real system. However, these calculations were based on analytical model systems, and they did not deal the real carrier excitations.

More recently, Yamaguchi et al. [30] evaluated the carrier dynamics using first-principles calculations. In their calculations, they calculated the electron excitation

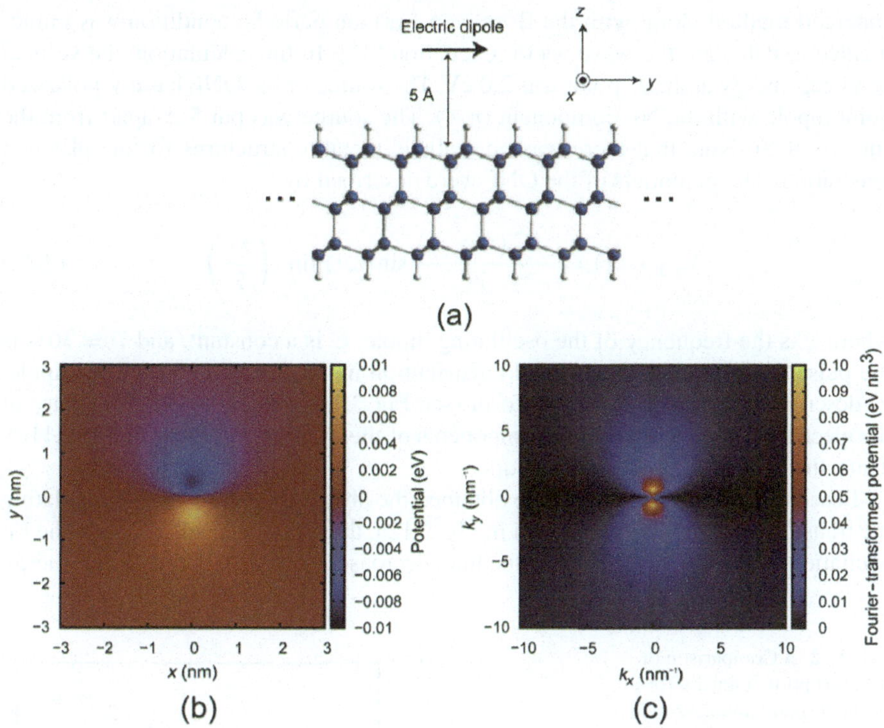

Fig. 2.11 **a** Theoretical model consisting of a Si bilayer and the electric dipole is measured at a distance of 5 Å above Si surface. **b** Potential of the ONF induced by the electric dipole shown in **a**. **c** The Fourier transform of **b**. **d** Comparison of the absorption intensity for far-field (blue) and near-field (red) excitations. **e** Absorption intensity as a function of the variation in the wave vector (Δk) for the excitation at 1.6 eV. These figures are reproduced from [33] under the terms of the Creative Commons Attribution 4.0 International license

by considering the indirect interband transition between states with different Bloch wave numbers. Instead of using the Bloch condition, which cannot describe the indirect interband transitions, they implemented the Born–von Karman boundary condition [31, 32]. Further, they calculated the time-dependent ONF excitation using one-dimensional Kronig-Penny (1D-KP) model (Fig. 2.9a, b). They considered the dispersion relation with an indirect band gap (Fig. 2.9c).

These results show that the absorption spectra for FF excitation decrease at the band gap energy (E_d), while the absorption spectra for NF excitation decrease at the band gap energy (E_g) ($< E_d$), meaning that the direct excitation was realized in the indirect band gap system. Furthermore, it was found that the NF component of the dipole radiation (r^{-3}) is the origin of NF excitation [1] (refer to (2.4)).

We investigated a more detailed system of the ONF excitation in realistic three-dimensional Si systems (Fig. 2.10a) [33]. Here, we used a real-time and real-space grid-based time-dependent density functional theory (DFT) approach [34–36]. A supercell method along with the Born–von Karman periodic condition was implemented to calculate the wave vector excitation [31]. In the calculation, the vertical band gap energy at the Γ point was 2.0 eV. The source of the ONF has a y-polarized point dipole with the NF component (r^{-3}). The source was put 5 Å apart from the surface of Si. Note that there was no realistic metallic structures for the plasmon generation. The potentials of the ONF were described by

$$V_{\text{near}}(r,t) = -\frac{C\left(y - y_{\text{p}}\right)}{r^3}\sin(\omega t)\sin^2\left(\frac{\pi t}{T}\right). \tag{2.32}$$

where ω is the frequency of the oscillating dipole, C is a constant, and $T = 30\,\text{fs}$ is the pulse duration. The electric field distribution induced by the oscillating dipole, as described by (2.32), is non-uniform (see Fig. 2.11b). Such a non-uniformity of the electric field generates large components of the Fourier domain (see Fig. 2.11c). Thus, the ONF has large wave vectors.

Based on these observations, we obtained the absorption intensity by calculating the transition probability. As shown in Fig. 2.12a, the absorption intensity for the FF excitation (blue solids circles and blue line) decreases ~2.0 eV (the band gap energy

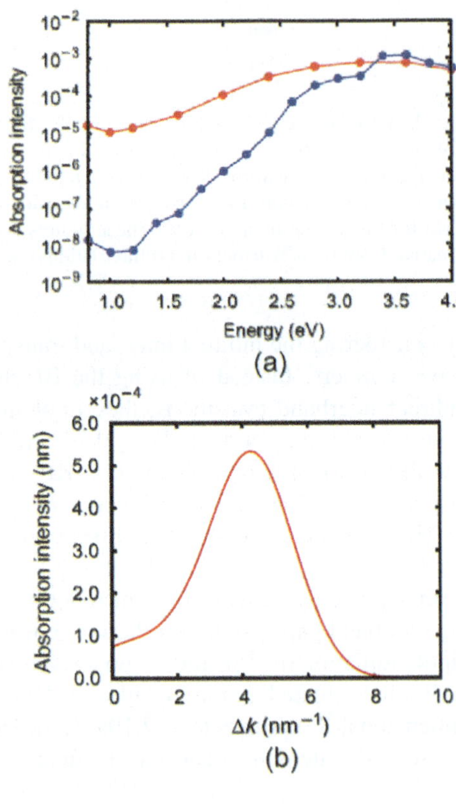

Fig. 2.12 **a** Comparison of the absorption intensity for far-field (blue) and near-field (red) excitations. **b** Absorption intensity as a function of the variation in the wave vector (Δk) for the excitation at 1.6 eV. These figures are reproduced from [33] under the terms of the Creative Commons Attribution 4.0 International license

at the Γ point). However, the NF excitation resulted in a redshift of the absorption spectra and large absorption intensity (red solid circles and red line in Fig. 2.12a). In addition, we confirmed direct optical transitions between different wave vectors. Figure 2.12b shows the absorption intensity as a function of the variation in wave vector for the electronic excitation between the second-highest valence band and the lowest conduction band in the ONF excitation at 1.6 eV. Here, the absorption intensity takes maximum at $\Delta k \sim 4\ \mathrm{nm}^{-1}$. This was the direct evidence that the ONF induces the wave vector excitation in the indirect band gap system. In other words, the direct ONF excitation does not require phonon assistance or plasmon resonance, and the direct ONF excitation can be realized with photons only.

References

1. *Classical Electrodynamics* (Wiley, Hoboken, 1962)
2. T. Yatsui, M. Yamaguchi, K. Nobusada, Nano-scale chemical reactions based on non-uniform optical near-fields and their applications. Prog. Q. Electron. **55**, 166–194 (2017)
3. R.J. Levis, G.M. Menkir, H. Rabitz, Selective bond dissociation and rearrangement with optimally tailored, strong-field laser pulses. Science **292**(5517), 709–713 (2001)
4. S. Kim, J. Jin, Y.-J. Kim, I.-Y. Park, Y. Kim, S.-W. Kim, High-harmonic generation by resonant plasmon field enhancement. Nature **453**(7196), 757–760 (2008)
5. M. Yamaguchi, K. Nobusada, Photodissociation path in H_2^+ induced by nonuniform optical near fields: two-step excitation via vibrationally excited states. Phys. Rev. A **93**(2), 023416 (2016)
6. P.H. Bucksbaum, A. Zavriyev, H.G. Muller, Softening of the H_2^+ molecular bond in intense laser fields. Phys. Rev. Lett. **64**(16), 1883–1886 (1990)
7. A. Giusti-Suzor, X. He, O. Atabek, F.H. Mies, Above-threshold dissociation of H_2^+ in intense laser fields. Phys. Rev. Lett. **64**(5), 515–518 (1990)
8. A. Giusti-Suzor, F.H. Mies, Vibrational trapping and suppression of dissociation in intense laser fields. Phys. Rev. Lett. **68**(26), 3869–3872 (1992)
9. K. Sändig, H. Figger, T.W. Hänsch, Dissociation dynamics of H_2^+ in intense laser fields: investigation of photofragments from single vibrational levels. Phys. Rev. Lett. **85**(23), 4876–4879 (2000)
10. P.M. Paul, E.S. Toma, P. Breger, G. Mullot, F. Augé, Observation of a train of attosecond pulses from high harmonic generation. Science **292**(5522), 1686–1689 (2001)
11. K.C. Kulander, F.H. Mies, K.J. Schafer, Model for studies of laser-induced nonlinear processes in molecules. Phys. Rev. A (1996)
12. N. Watanabe, M. Tsukada, Fast and stable method for simulating quantum electron dynamics. Phys. Rev. E **62**(2), 2914–2923 (2000)
13. J.H. Posthumus, The dynamics of small molecules in intense laser fields. Rep. Prog. Phys. **67**(5), 623–665 (2004)
14. J.R. Zurita-Sánchez, L. Novotny, Multipolar interband absorption in a semiconductor quantum dot. I. Electric quadrupole enhancement. J. Opt. Soc. Am. B **19**, 1355 (2002)
15. P.K. Jain, D. Ghosh, R. Baer, E. Rabani, Near-field manipulation of spectroscopic selection rules on the nanoscale. Proc. Natl. Acad. Sci. U.S.A. **109**, 8016 (2012)
16. A.M. Kern, O.J.F. Martin, Strong enhancement of forbidden atomic transitions using plasmonic nanostructures. Phys. Rev. A **85**(2), 022501 (2012)
17. R. Filter, S. Mühlig, T. Eichelkraut, C. Rockstuhl, F. Lederer, Controlling the dynamics of quantum mechanical systems sustaining dipole-forbidden transitions via optical nanoantennas. Phys. Rev. B **86**(3), 035404 (2012)

18. S. Tojo, M. Hasuo, T. Fujimoto, Absorption enhancement of an electric quadrupole transition of cesium atoms in an evanescent field. Phys. Rev. Lett. **92**(5), 053001 (2004)
19. M. Yamaguchi, K. Nobusada, T. Kawazoe, T. Yatsui, Two-photon absorption induced by electric field gradient of optical near-field and its application to photolithography. Appl. Phys. Lett. **106**(19), 191103 (2015)
20. M. Yamaguchi, K. Nobusada, T. Yatsui, Nonlinear optical response induced by a second-harmonic electric-field component concomitant with optical near-field excitation. Phys. Rev. A **92**, 043809 (2015)
21. W. Robert, R.W. Boyd, *Nonlinear Optics* (Academic, Cambridge, 2008)
22. L. Novotny, B. Hecht, *Principle of Nano-Optics* (Cambridge University Press, Cambridge, 2006)
23. P.K. Jain, D. Ghosh, R. Baer, E. Rabani, A.P. Alivisatos, Near-field manipulation of spectroscopic selection rules on the nanoscale. Proc. Nat. Acad. Sci. USA **109**, 8016 (2012)
24. D.M. Schaadt, B. Feng, E.T. Yu, Enhanced semiconductor optical absorption via surface plasmon excitation in metal nanoparticles. Appl. Phys. Lett. **86**(6), 063106 (2005)
25. K. Nakayama, K. Tanabe, H.A. Atwater, Plasmonic nanoparticle enhanced light absorption in GaAs solar cells. Appl. Phys. Lett. **93**(12), 121904 (2008)
26. C.F. Bohren, D.R. Huffman, *Absorption and Scattering of Light by Small Particles* (Wiley, Hoboken, 1998)
27. M. Kirkengen, J. Bergli, Y.M. Galperin, Direct generation of charge carriers in c-Si solar cells due to embedded nanoparticles. Journal of Applied Physics **102**(9), 093713 (2007)
28. J. Jung, M.L. Trolle, K. Pedersen, T.G. Pedersen, Indirect near-field absorption mediated by localized surface plasmons. Phys. Rev. B **84**, 165447 (2011)
29. T. Yatsui, S. Okada, T. Takemori, T.S.K. Saichi, T. Ogamoto, S. Chiashi, S. Maruyama, M. Noda, K. Yabana, K. Iida, K. Nobusada, Enhanced photo-sensitivity in a si photodetector using a near-field assisted excitation. Commun. Phys. **2**, 62 (2019)
30. M. Yamaguchi, K. Nobusada, Indirect interband transition induced by optical near fields with large wave numbers. Phys. Rev. B **93**, 195111 (2016)
31. R.M. Martin, *Electronic Structure: Basic Theory and Practical Methods* (Cambridge University Press, Cambridge, 2004)
32. N.W. Ashcroft, N.D. Mermin, *Solid State Physics* (Hold, Rinehart, and Winston, 1976)
33. M. Noda, K. Iida, M. Yamaguchi, T. Yatsui, K. Nobusada, Direct wave-vector excitation in an indirect-band-gap semiconductor of silicon with an optical near-field. Phys. Rev. Applied **11**, 044053 (2019)
34. M. Noda, K. Ishimura, K. Nobusada, K. Yabana, T. Boku, Massively-parallel electron dynamics calculations in real-time and real-space: toward applications to nanostructures of more than ten-nanometers in size. J. Comput. Phys. **265**, 145–155 (2014)
35. M. Noda, M. Yamaguchi, K. Nobusada, Second harmonic excitation of acetylene by the optical near field generated in a porous material. J. Phys. Chem. C **121**(21), 11687–11692 (2017)
36. K. Iida, M. Noda, K. Nobusada, Development of theoretical approach for describing electronic properties of hetero-interface systems under applied bias voltage. J. Chem. Phys. **146**(8), 084706 (2017)

Chapter 3
Ultra-flat Surface Using a Near-Field Etching

3.1 Introduction

Based on the properties of the ONF, it is expected to realize unique fabrication technique by the use of the ONF, using unique properties including the multi-step excitation via dipole-forbidden intermediate state of the molecules, second harmonic generation (SHG), wave vector excitation. Since the ONF localize at the vicinity of the nanoscale structure, we considered that we can realize selective chemical reaction at the nanoscale structure. One of the applications, we developed the near-field etching [1].

3.2 Realization of an Ultra-flat Surface by the Near-Field Etching

The process of the near-field etching is explained as follows. At the nanoscale, even for the flattened substrate prepared by a chemical mechanical polishing (CMP), there still remain the nanoscale protrusions in atomic scale as shown in Fig. 3.1. Against these samples, when the light incident to the substrate, the optical near-field is generated at the apex of the protrusions. In which, if the photon energy of the light is lower than that of absorption edge energy of the molecules, the light is not absorbed by the molecules in the gas phase. Therefore, the molecules are not dissociated.

On the contrary, if the molecules come into the area where the ONF generated, the molecule can be excited owing to the near-field excitation including multi-step excitation, or SHG. Consequently, the molecules are dissociated in the vicinity of the protrusions. Then, since the dissociated molecules have high activity of the chemical reaction, the protrusions are etched selectively. Finally, we could obtain the ultra-flat surface with atomic scale Fig. 3.1. After the disappear of the protrusions, the etching process will be stopped because the ONF disappear.

T. Yatsui, *Nanophotonic Chemical Reactions*, Nano-Optics and Nanophotonics,
https://doi.org/10.1007/978-3-030-42843-3_3

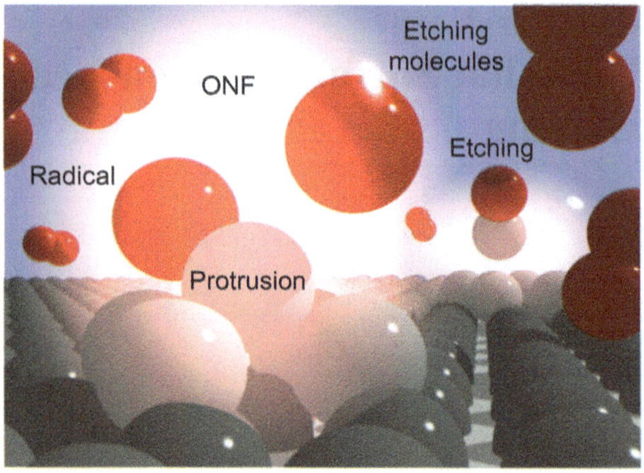

Fig. 3.1 Schematic of the near-field etching. This figure has been reproduced from [2] under the terms of the creative commons attribution 4.0 international license

More detailed of the near-field etching was explained by the examples of the cases using glass substrate and the etching molecule of Cl_2 [1]. Since the absorption edge wavelength of Cl_2 is around 400 nm [3, 4], the wavelength for the near-field etching was set to 532 nm, so that it results in the selective dissociation of Cl_2 only in the vicinity of the protrusions.

As for the glass substrate, we used the synthetic silica substrates with an OH group concentration of less than 1 ppm [5]. Figure 3.2a shows the typical atomic force microscopic (AFM) image of the before etching, in which several protrusions and scratches due to the CMP were observed. However, after 60 min near-field etching, they were diapered as shown in Fig. 3.2b.

Figure 3.3 shows the time dependence of the surface roughness, $\bar{R}_a$, of the synthetic silica substrate. The R_a value is an arithmetic average of the absolute values of the surface height deviations measured from a best-fit plane and is given by

$$R_a = \frac{1}{l}\int_0^l |f(x)|\mathrm{d}x \cong \frac{1}{n}\sum_{i=1}^{n} |f(x_i)| \tag{3.1}$$

The R_a value was measured by AFM with a scanning area of $10\,\mu\text{m} \times 10\,\mu\text{m}$. Therefore, we could not evaluate all over the substrate. Thus, to evaluate the R_a value with a wide area, we obtained averaged R_a values at different 9 points, $\bar{R}_a$, in which each point was separated by $100\,\mu\text{m}$ with 3×3 matrix. As shown in Fig. 3.3, the roughness decreased as the etching time increased. We obtained the ultra-flat surface as small as $\bar{R}_a = 0.15$ nm. The value of the surface roughness was saturated after 60 min. This might be caused by the measurement limitation of the AFM, because the value is comparable to the size of the atomic separation.

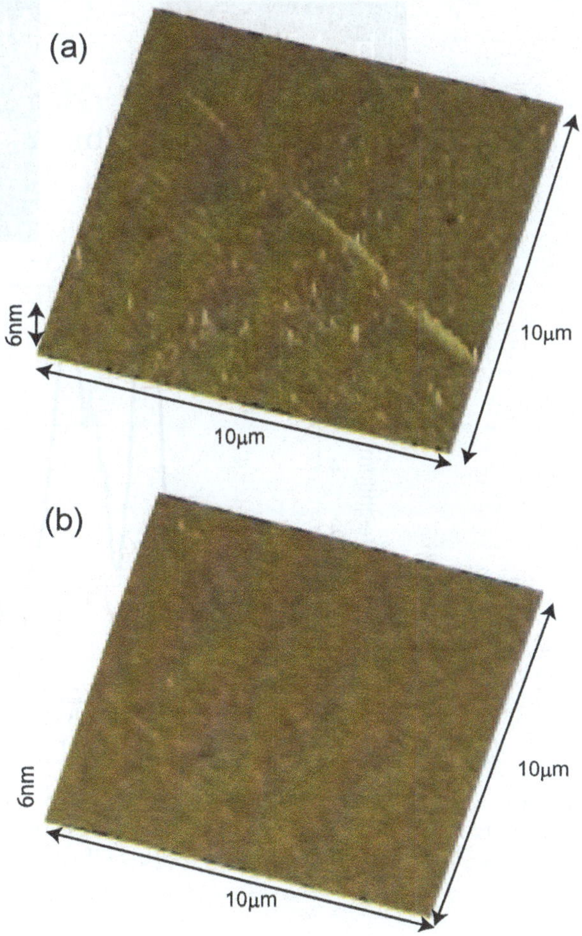

Fig. 3.2 Typical AFM images of the synthetic silica substrate: **a** before etching, **b** after etching (etching time was 60 min). These figures are reprinted with permission from [1], Copyright © 2008 Springer Nature

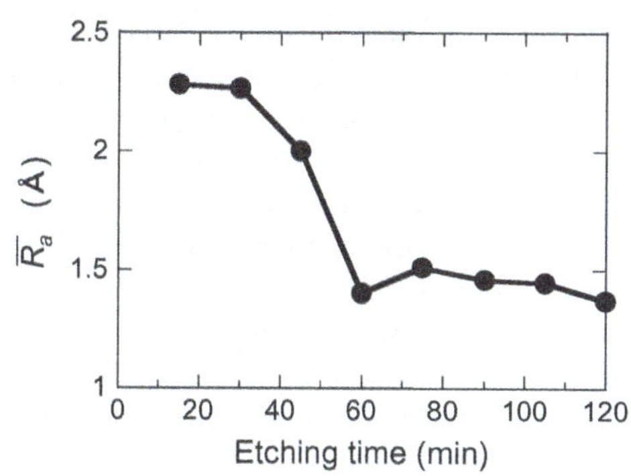

Fig. 3.3 Etching time dependence of R_a of the synthetic silica substrate. The figure is reprinted with permission from [1], Copyright © 2008 Springer Nature

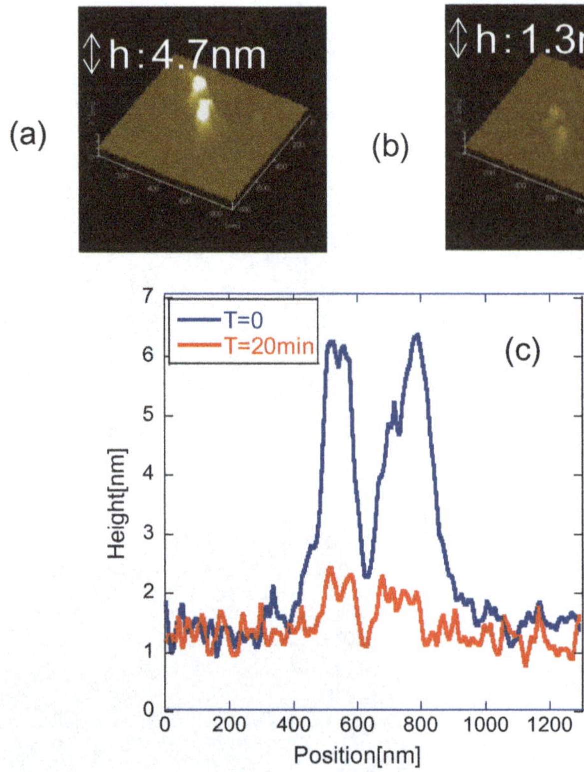

Fig. 3.4 AFM images **a** before (T = 0 min) and **b** after 20 min (T = 20 min) the near-field etching of the LTEM substrate. **c** Cross-sectional profiles along the protrusions at the same position. These figures are reprinted from [6], Copyright © 2013 SPIE

To confirm that the near-field etching selectively etch away the protrusions, we compared the same position between before and after the near-field etching [6]. As for this evaluation, we used the low-thermal expansion material (LTEM) substrates made of Titania-doped fused silica [7]. The position reproducibility of the AFM system is not high, and the area of AFM is limited in several ten μm. In addition, because we need to detach the substrate after the AFM measurement and put it again after the near-field etching to measure the AFM image, it is difficult to measure the same positions. Thus, we prepared the mark on the substrate using a focused ion beam. We prepared the mark several μm away from the pits or bumps after the AFM measurement in prior to the near-field etching. Figure 3.4a, b shows the before and after the near-field etching at the same position. From these results, the protrusions were selectively etched and the height decreased (see Fig. 3.4).

Using the same substrate, we investigated the morphology change of the pit. Figure 3.5a shows the AFM image in which both the pit and bump include before etching. As shown in Fig. 3.5b, c, the depth of the pit decreased, as well as the height

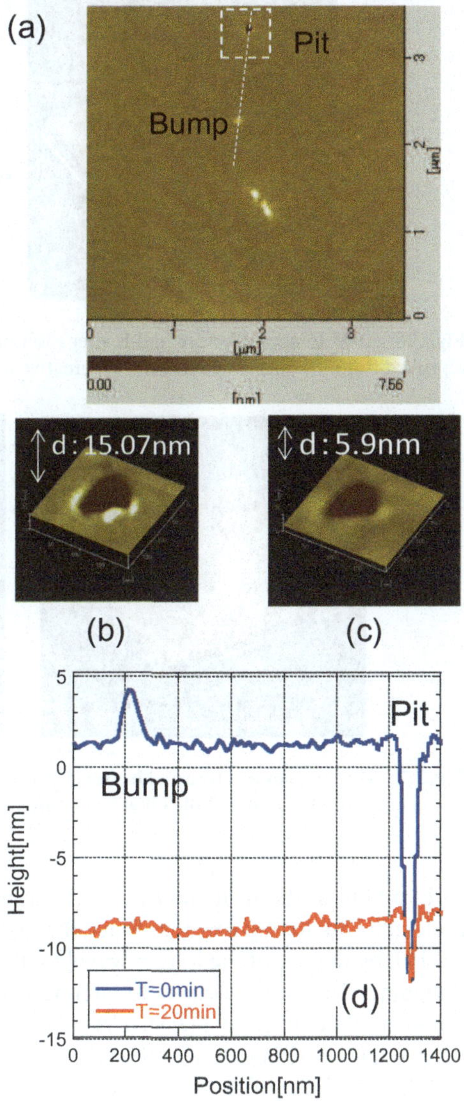

Fig. 3.5 AFM images **a** before (T = 0 min) etching of the LTEM substrate. Magnified AFM images at the pit (dashed square in **a**) **b** before and **c** after 20 min (T = 20 min) the near-field etching. **d** Cross-sectional profiles along the white dashed line in **a**. These figures are reprinted from [6], Copyright © 2013 SPIE

of the bump decreased. From the cross-sectional profile along the pit and bump (see Fig. 3.5d), the flat area also etched, and finally we obtained flat surface over the pits and bumps.

The near-field etching is based on the photochemical reaction, and it can be applicable to other molecules as an etching gas and/or other materials to be etched. We performed the near-field etching on (001) GaN substrate [8] using the Cl_2 gas and 532 nm light. The AFM images taken before (Fig. 3.6a) and after 30 min (Fig. 3.6b) of etching show that R_a decreased from 0.23 to 0.14 nm.

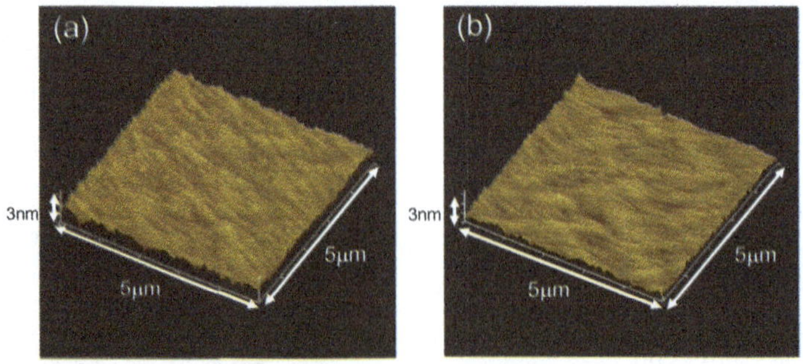

Fig. 3.6 AFM images **a** before and **b** after (30 min) etching of GaN substrate. These figures are reprinted from [8] under the terms of the creative commons attribution license

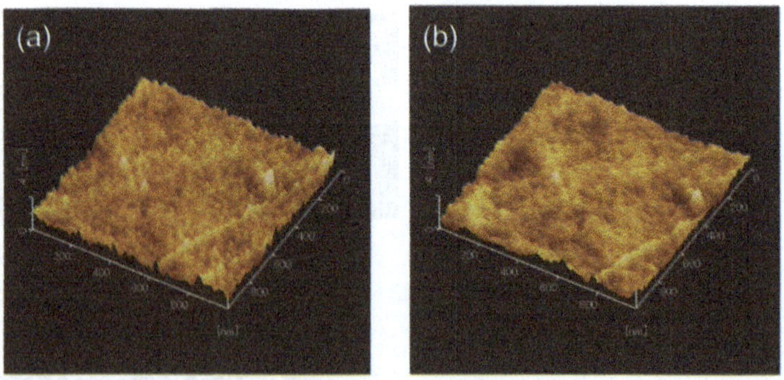

Fig. 3.7 AFM images **a** before and **b** after (120 min) etching of PMMA. These figures are reprinted from [11] under the terms of the creative commons attribution license

In addition, the near-field etching can be applicable to the use in other etching molecules, including oxygen. Since the O_2 dissociation energy is 5.12 eV (242 nm) [9, 10], by the use of the longer wavelength than that, it is expected to realize similar flattening. To test the near-field etching using oxygen, we used plastic substrate of polymethyl methacrylate (PMMA) [11]. As for the light source, a He-Cd laser with a wavelength of 325 nm longer than the oxygen absorption edge was used. As shown in Fig. 3.7, the value of R_a decreased from 0.334 nm (Fig. 3.7a) to 0.265 nm (Fig. 3.7b) after 120 min. near-field etching.

In order to confirm that the selective etching of the protrusions where the ONF generated by the near-field etching, we evaluated the wavelength dependence of the surface morphology. For this evaluation, instead of the use of R_a value (Fig. 3.8a), we used the standard deviation of the height difference function. This is because that the R_a value is the average value of the absolute surface height deviations from the average height of the surface. As for the value, we introduced the new parameter of the standard deviation of the height difference function which is given by following (see Fig. 3.8b)

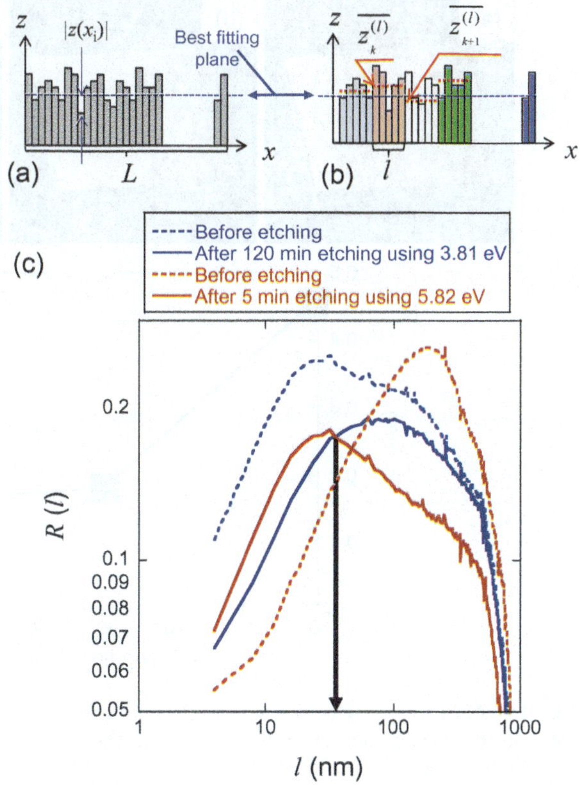

Fig. 3.8 Schematic of the value of **a** R_a (determined by (3.1)) and **b** $R(l)$ (determined by (3.2)). **c** Comparison of the value $R(l)$. These figures are reprinted from [11] under the terms of the creative commons attribution license

$$R(l) = \sqrt{\left\langle \left(\overline{z_{k+1}^{(l)}} - \overline{z_{k}^{(l)}} \right)^2 /2 \right\rangle} \tag{3.2}$$

In this expression, l value is the scale size, z is the height from the best-fitting plane, and $\overline{z_k^{(l)}}$ is the average z value of the scale [12]. This $R(l)$ can be used to determine the contributions of the surface roughness values (at different length scales) to the overall surface roughness. As this value indicates the height difference of the next scale, it is known as a two-sample variance. Using this value, we compared between the near-field etching using 325-nm (3.81 eV) light (longer than the absorption edge of oxygen) and the far-field etching using 213-nm (5.82 eV) light (shorter than the absorption edge of oxygen) (see Fig. 3.8c). From the comparison, it is found that the near-field etching selectively etched the scale less than 100 nm, in contrast, the far-field etching etched the scale larger than 100 nm. This result confirmed that the near-field etching etched the smaller scale selectively.

Similar reduction of the surface roughness R_a and change of the $R(l)$ were observed in the diamond substrate. Figure 3.9 shows the AFM images of the diamond

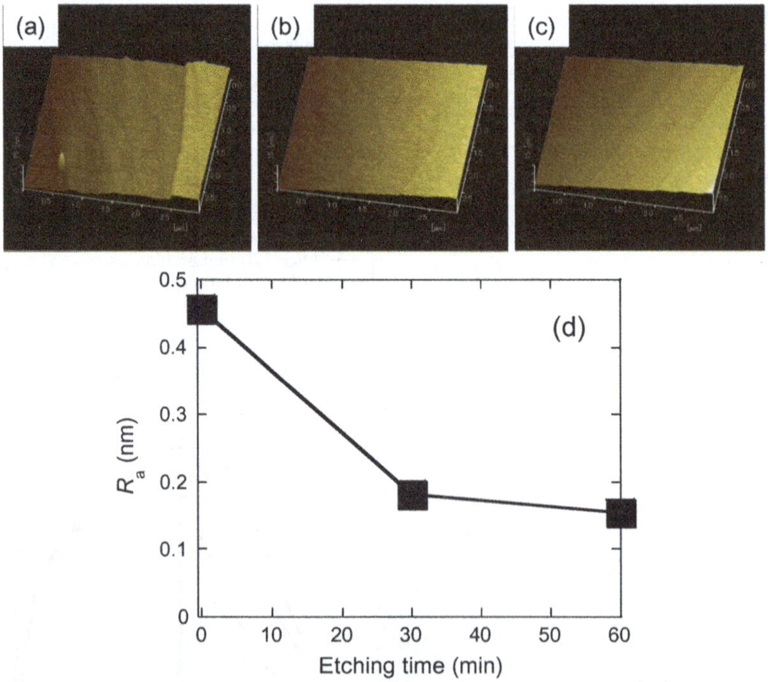

Fig. 3.9 AFM images **a** before, **b** after 30 min, and **c** after 60 min etching of diamond (111) substrate. **d** Time dependence of the value of R_a. These figures are reprinted with permission from [13], Copyright © 2012 IOP Publishing

Fig. 3.10 Schematic of the value of **a** R_a (determined by (3.1)) and **b** $R(l)$ (determined by (3.2)). **c** Comparison of the value $R(l)$. These figures are reprinted with permission from [13], Copyright © 2012 IOP Publishing

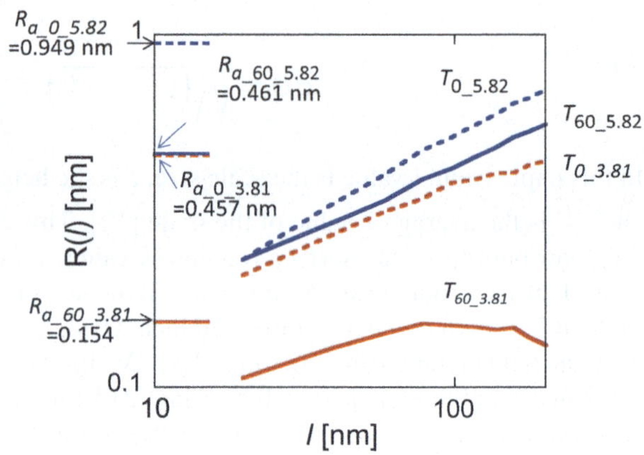

(111) substrate. By the 325-nm light irradiation, the value of the surface roughness R_a decreased from 0.457 nm (Fig. 3.9a) to 0.154 nm (Fig. 3.9c) with 60 min.

To check the selective etching by the near-field etching, the $R(l)$ value was evaluated. The $R(l)$ values were compared between that of the near-field etching using

325-nm light (3.81 eV) and that of the far-field etching using 213-nm light (5.81 eV). As shown in Fig. 3.10, for the far-field etching the component of the small l remained the same and the component of large l more than 100 nm decreased. In contrast, for the near-field etching the component of the small l decreased.

3.3 Near-Field Wet Etching

Since the near-field etching process utilize the photochemical reaction, it can be applicable to the usage of the molecules in the liquid phase [14]. To confirm this, a 25 wt.% calcium hypochlorous acid ($Ca(ClO)_2$) was used as a source ion for wet etching (see Fig. 3.11).

We performed the near-field wet etching using 532 nm light. Figure 3.12a–c shows the time dependence of AFM images, indicating the surface roughness was decreased by the near-field etching. Figure 3.12d shows the comparison of the time dependence of the surface roughness values between the near-field wet and dry etching. The near-field dry etching was performed using 532 nm light and Cl_2 gas with the partial pressure of 200 Pa. Although the initial surface roughness is not the same, the similar reduction of the surface roughness value as the near-field wet etching was observed.

As is described in the [14], the mean free paths (MFP) of Cl_2 and ClO^- were estimated. The gas-phase MFP of Cl_2 is on the order of $10\,\mu m$ ($\sim \frac{RT}{\sqrt{2}N_A \pi d^2 P}$ [15], where $R = 8.31$ J $K^{-1}mol^{-1}$, $T = 300$ K, $N_A = 6.02 \times 10^{23}$ mol^{-1}, $d = 200 \times 10^{-12}$ m (the diameter of a Cl_2 molecule), and $P = 200$ Pa). In contrast, the MFP of ions in solution is on the order of 10 pm [16], which is 10^{-7} times smaller than Cl_2 gas. Such a low MFP value for ions in solution resulted in the etching near the

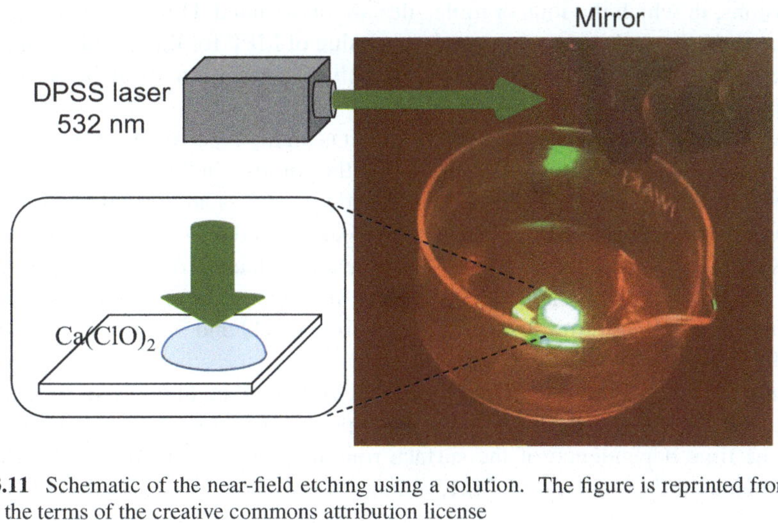

Fig. 3.11 Schematic of the near-field etching using a solution. The figure is reprinted from [14] under the terms of the creative commons attribution license

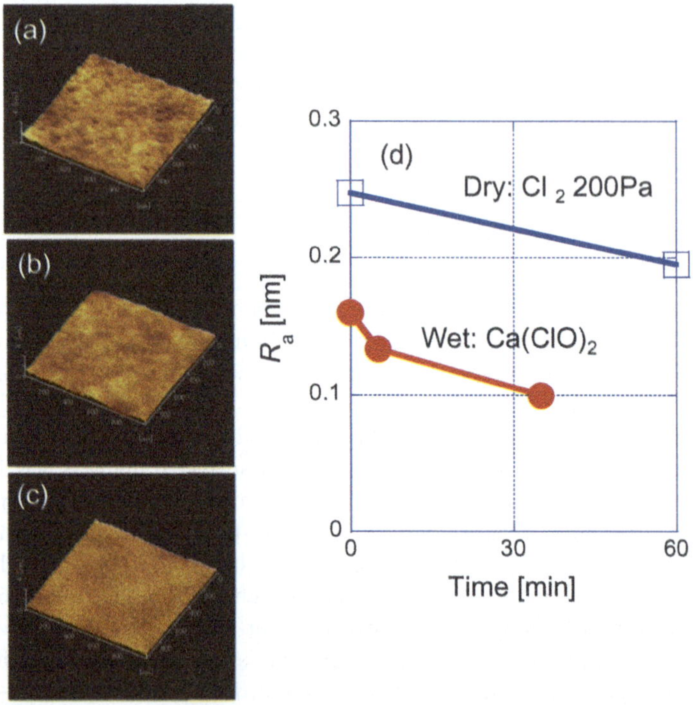

Fig. 3.12 AFM images of **a** before, **b** after 5 min, and **c** after 35 min near-field etching. **d** Comparison of the time dependence of the surface roughness R_a between the near-field dry and wet etching. Open squares: the near-field dry etching. Closed circles: the near-field wet etching. These figures are reprinted from [14] under the terms of the creative commons attribution license

protrusions, in which the ions or molecules are dissociated. However, in the case of the near-field dry etching, owing to a larger value of MFP for the gas phase than the scanned area the dissociated atoms can react only when they remain at the protrusions where the ONF generated.

The chlorine is one of halogens atoms, and it is highly reactive. Therefore, we also tested one of the other halogen, iodine. As for the solution including iodine, we used the mixture of I_2 and KI. I_2–KI solution is widely used as an etchant for gold [17]. Although iodine solution has an absorption peak around 550 nm, when it is mixed with KI, I_2 dissolved and it forms I_3^-. Then, the color change from dark red of iodine solution to light yellow of I_2–KI solution [18]. Figure 3.13 shows the absorption spectrum of I_2–KI solution, in which peaks around 290 and 350 nm correspond to that of I_3^-.

Using the I_2–KI solution, we performed the near-field etching. First, we checked I_2–KI solution does not affect the surface roughness. The open squares in Fig. 3.14a show the time dependence of the surface roughness R_a. From the results, the R_a value remained the same around 0.15 nm. As a light source, we used He-Cd laser

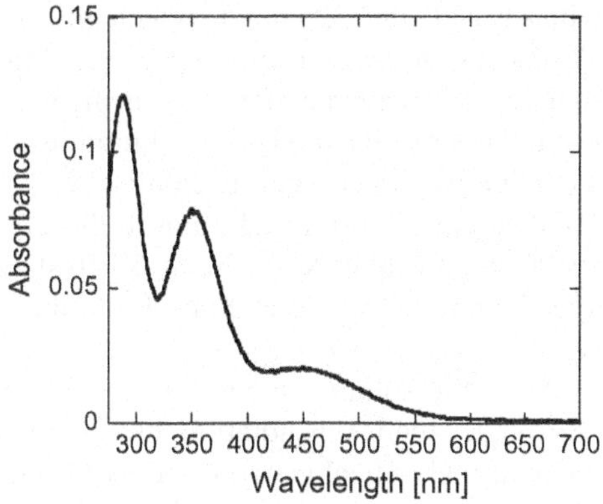

Fig. 3.13 Absorption spectrum of I_2–KI solution. I_2 concentration was less than 1%

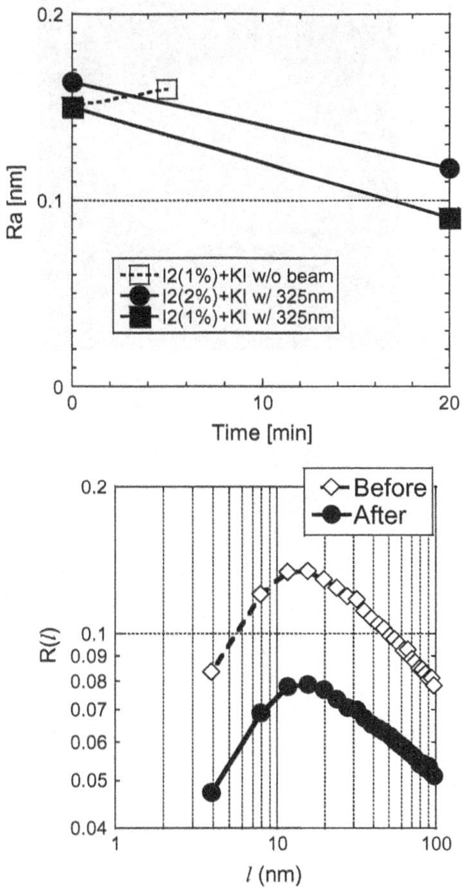

Fig. 3.14 **a** Time dependence of the surface roughness R_a using I_2–KI solution. Dashed line and open squares: without laser irradiation. I_2 concentration was less than 1%. Solid line and closed circles: with laser irradiation. I_2 concentration was less than 2%. Solid line and closed squares: with laser irradiation. I_2 concentration was less than 1%. **b** Time dependence of the value $R(l)$. I_2 concentration was less than 1%

with a wavelength 325 nm, in which the power was 50 mW and a spot size was 1 mm. The closed circles and squares in Fig. 3.14a show the time dependence of the surface roughness R_a with the irradiation using I_2 concentrations of 2% and 1%, respectively. From these results, the I_2–KI solution resulted in the reduction of the R_a value. The R_a value was decreased as small as 0.090 nm in the case of I_2 concentration of 1% (see Fig. 3.14a). In addition to the evaluation of the surface roughness R_a, we evaluated the value $R(l)$. Figure 3.14b shows the calculated $R(l)$ values before and after 20 min near-field etching. From the results, the $R(l)$ values decreased at all l value.

As of the solution for the near-field etching, we tested the hydrogen peroxide (H_2O_2). By the UV irradiation through hydrogen peroxide solution, it is known that the hydroxyl radical generates as expressed by the following [19, 20]

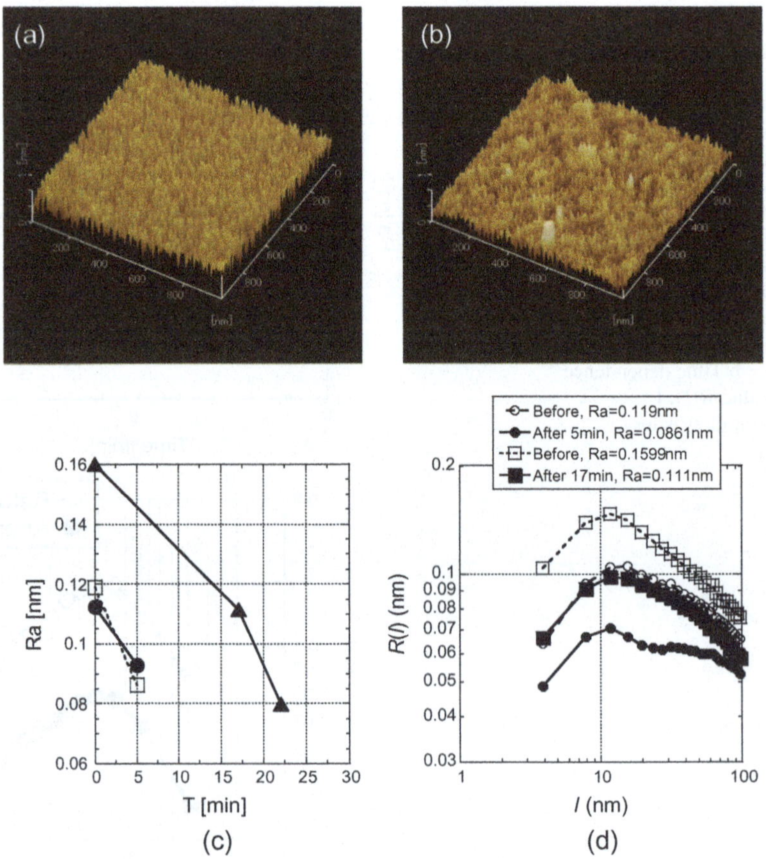

Fig. 3.15 AFM images **a** before and **b** after (22 min) near-field etching by using H_2O_2. The surface roughness R_a are 0.160 and 0.080 nm, respectively. **c** Time dependence of the surface roughness R_a. **d** Time dependence of the value $R(l)$

$$H_2O_2 + h\nu\,(< 248\,\text{nm}) \rightarrow 2\text{OH}. \tag{3.3}$$

The hydroxyl radicals are highly reactive, and it is expected to react the surface. We performed the near-field wet etching using 213 nm light (5.6 mJ, 20 Hz) with the hydrogen peroxide. Figure 3.15a, b shows AFM images of before and after the near-field etching, respectively.

The surface roughness R_a decreased from 0.160 to 0.080 nm after 22-min near-field etching. Similar etching was done on the same substrate with initial surface roughness R_a of 0.119 and 0.112 nm. The surface roughness R_a decreased to 0.086 and 0.093 nm, respectively. We confirmed the use of the hydrogen peroxide could realize the super flat substrate with the sub Å scale. We also evaluated the $R(l)$ values from AFM images. Figure 3.15d shows the calculated $R(l)$, indicating that the near-field wet etching resulted in the decrease of $R(l)$ values over the all scale l.

3.4 High Speed and Large Area Flattening Using the Near-Field Etching

We have confirmed the near-field etching can realize the ultra-flat surface. In addition, we apply this technique to high-speed and large area flattening. As for the sample, we used the substrate for 2.5 in. HDD made of glass. Figure 3.16a shows the schematic of the substrate. The edge of the substrate was cut at 45° to avoid the damage during the handling, which is called as chamfer.

The initial surface roughness R_a of the chamfer is large more than 100 nm (see Fig. 3.16b). Thus, the chamfer is polished using CMP, and the surface roughness R_a decreases as small as 5 nm (see Fig. 3.16c).

We performed the near-field etching on the chamfer of the hard disk substrate in the dry etching chamber (Fig. 3.17), so that we could exclude the CMP process for hard disk substrate.

We used a light source with a wavelength of 532 nm and Cl_2 as a gas source. We put the several substrates together (see Fig. 3.17d). The laser beam with a wavelength of 532 nm was focused as small as 0.2 mm in a diameter. The beam was scanned with a speed of 0.22 mm/s. The power density was estimated to be 42 kW/cm^2.

Figure 3.18 shows the AFM image at the chamfer after the near-field etching. The surface roughness R_a decreased as small as 5.0 nm. This value is comparable to that polished by CMP, as shown in Fig. 3.16c.

To realize higher throughput of the process, we performed the near-field etching in the outside the chamber (see Fig. 3.19). The chamfer was coated by the black ink. The laser beam was focused as small as 0.2 mm in diameter. The power density was estimated as 21 kW/cm^2. The beam was chopped with 2 kHz to avoid the heating of the substrate.

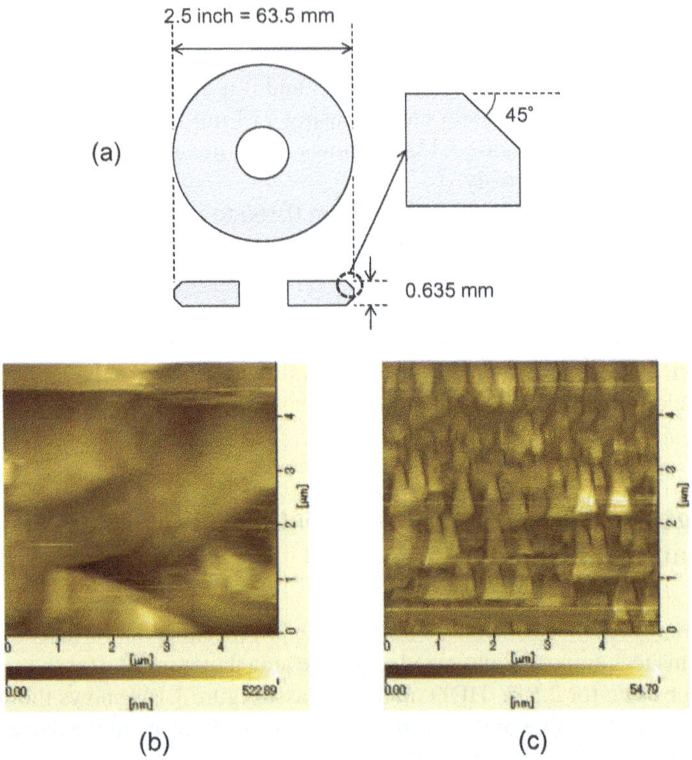

Fig. 3.16 **a** Schematic of the 2.5 in. hard disk substrate. Atomic force microscopic images (AFM) of **b** before and **c** CMP etching. The surface roughness R_a was **b** 132.7 and **c** 5.0 nm, respectively

Figure 3.19d shows the AFM image after the near-field etching. The surface roughness R_a decreased as small as 4.5 nm, which was smaller than that in chamber with Cl_2 gas. From these results, it is expected the near-field etching of the glass substrate can be performed without using Cl_2 gas in vacuum chamber. Furthermore, the etching speed was 7.2 s per substrate.

In addition to the above experiment, we performed the etching without the optical chopper (see Fig. 3.20a). As shown in Fig. 3.20b, the edge of the substrate where the light was focused was peeled after the light irradiation, which might be caused by the heating. Figure 3.20d shows the AFM image after this process on the peeled area, and the surface roughness R_a was found to be decreased as small as 0.99 nm.

We have tried the near-field etching on the recording surface of the 2.5 in. hard disk. To irradiate over the surface, we scanned the sample stages with rotation stage (7.3 s/rotations) and one-direction stage for radius direction (scanning speed of 0.02 mm/s between $r_1 = -8$ mm and $r_2 = -14$ mm). The hard disk sample was dipped in the H_2O_2 solution. We used the 213 nm light with a beam spot of 4 mm in diameter (see Fig. 3.21a). In prior to the near-field etching experiment, we checked the scanned area.

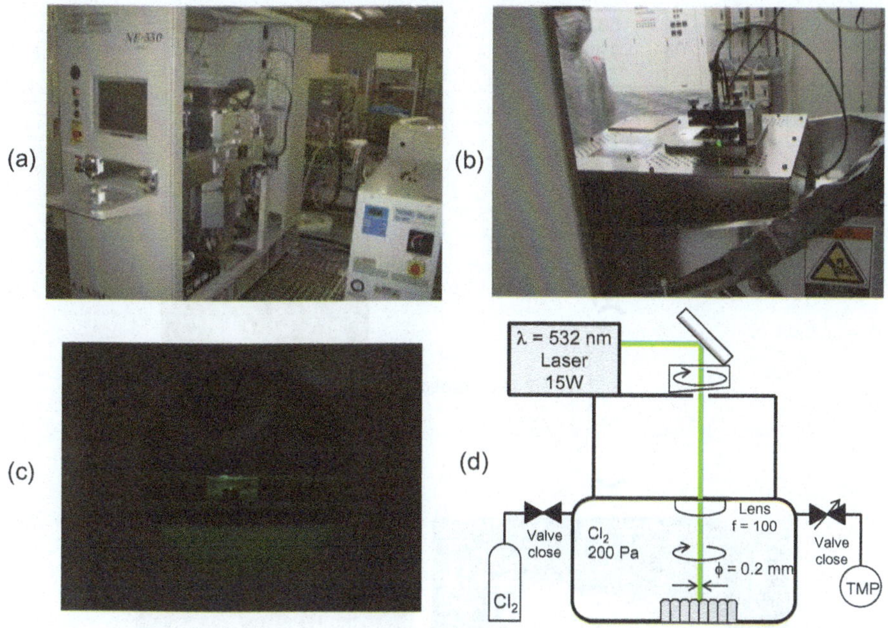

Fig. 3.17 Near-field etching. **a** Overview of the inductively coupled plasma (ICP) chamber for the near-field etching. **b** Laser introduced from the top view port of ICP chamber. **c** During the process, from the side view port. **d** Schematic of the near-field etching process in the ICP chamber

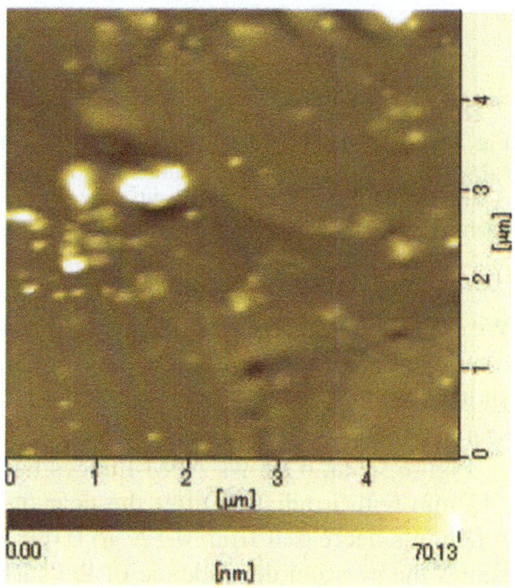

Fig. 3.18 AFM image at the chamfer after the near-field etching

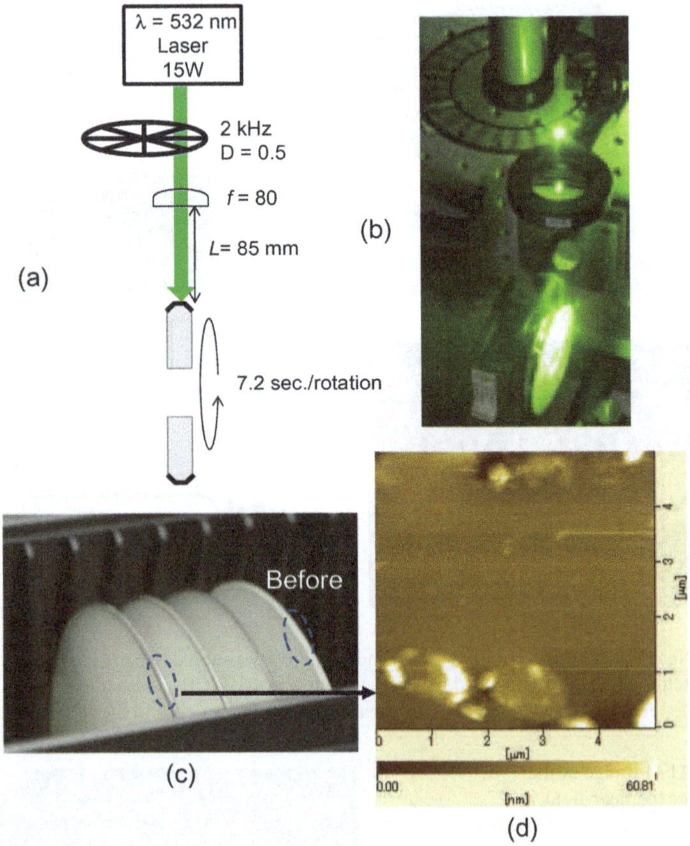

Fig. 3.19 Near-field etching in the outside the chamber. **a** Schematic of the near-field etching. The substrate was rotated at 7.2 s/rotation. **b** Photograph of **a**. **c** Photograph of the substrates. Right and right edges of them were before and after etching. **d** AFM image of the near-field etching at the chamfer

We used UVSCALE (FUJIFILM Corporation) which can visualize the UV light spot. The UVSCALE was put below the sample, however without solution. Figure 3.21b shows the photograph of the UVSCALE, in which the dark area corresponds the scanned area.

Figure 3.22a, b shows AFM images taken at –21 mm with no irradiation and at –13 mm with irradiation after the near-field etching. It was found that the surface roughness decreased from 0.127 to 0.098 after 3-h near-field etching. Figure 3.22c shows the position dependence of the surface roughness R_a. It was found that the value of the surface roughness clearly decreased from 0.12 to 0.10 nm.

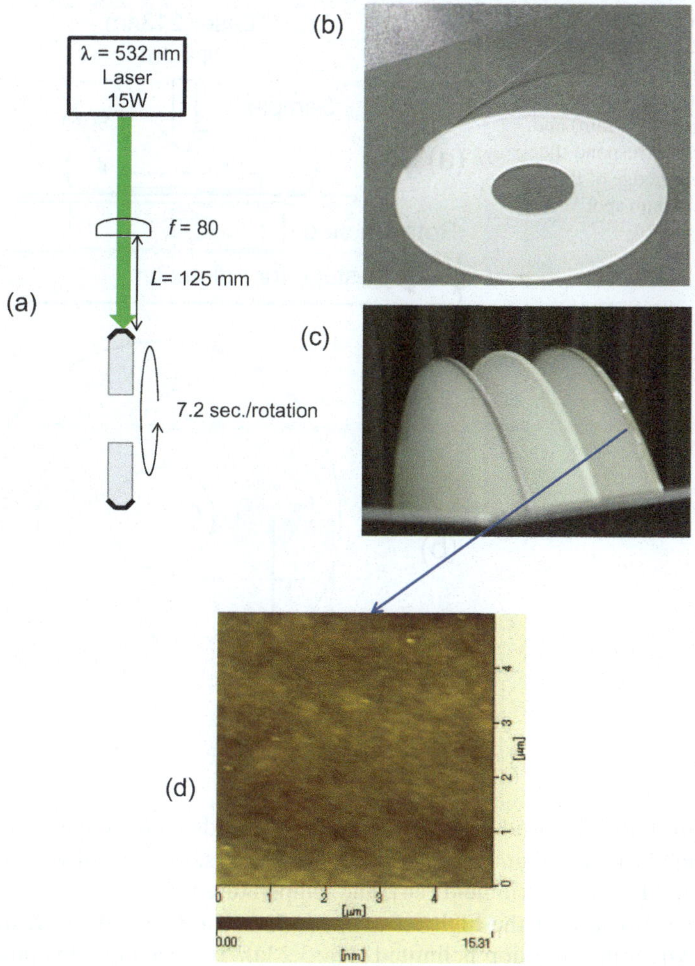

Fig. 3.20 Near-field etching in the outside the chamber. **a** Schematic of the near-field etching. The substrate was rotated at 7.2 s/rotation without the optical chopper. **b** Photograph of the etching. The substrate edge was peeled. **c** Photograph of the substrates. Right edge of them were after etching. **d** AFM image of the near-field etching at the chamfer

3.5 Realization of an Ultra-flat Surface of the Three-Dimensional Structures

The near-field etching is a non-contact method; therefore, it can be applicable to the non-flat substrate, i.e., the substrate with a three-dimensional (3D) structures [21].

As for the substrate with the 3D structures, we performed the near-field etching on the solid immersion lens (SIL) of a diamond. When the pair of nitrogen and vacancy (NV) is implanted in the diamond, the NV acts as a color center [22]. The emission

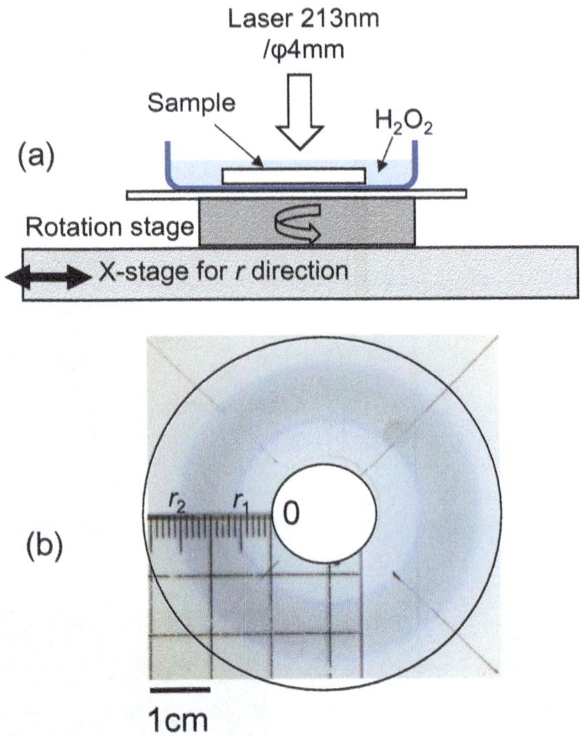

Fig. 3.21 **a** Schematic of the near-field etching on the recording surface of the hard disk. **b** Photograph of the UVSCALE. r_1 (− 8 mm) and r_2 (−14 mm) correspond the inner and outer edge of the center of the beam spot

intensity from the diamond NV center strongly depends on the environment. Thus, it is expected that the diamond NV could be a high sensitive sensors including a magnetic field [23], electric field [24], and temperature [25].

However, because of the high refractive index of a diamond, 2.42, the critical angle to extract the emission is limited (Fig. 3.23a). To overcome this problem, the solid immersion lens (SIL) structure is introduced (Fig. 3.23b). Since the diamond is very stable and hard material, it is very difficult to create the SIL structure. Thus, the SIL structure was fabricated by focused ion beam (FIB) (Fig. 3.23c) [26], and thus, the surface is not smooth.

To realize the smooth surface of the SIL, we performed the near-field etching. We used oxygen as for the etching gas and the He-Cd laser with a wavelength of 325 nm (10 mW), so that the wavelength is longer than the dissociation energy of the oxygen [9, 10]. Figure 3.24a, b shows the AFM images of before and after the near-field etching. As shown in those figures, it was found that the near-field etching could realize the smooth surface of the 3D structures.

We also applied to the lens array of the synthetic silica (Fig. 3.25). We used the lens array with the curvature radius of 100 mm, and the pitch of the lens of 1.015 mm. Using those substrates, we used Cl_2 as the gas source and 532 nm laser as the light for the near-field etching.

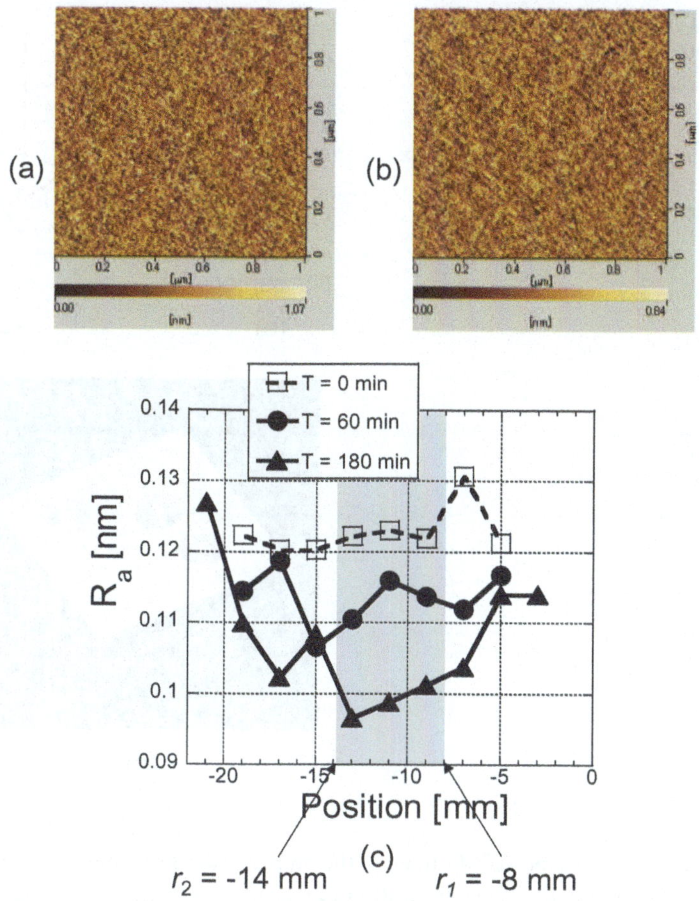

Fig. 3.22 AFM images at **a** −21 mm (outside the beam) and **b** −13 mm (inside the beam). Surface roughness R_a are 0.127 and 0.098 nm. **b** Position dependence of the surface roughness R_a. Open squares and dashed line: before etching, closed circles and solid line: after 60 m, and closed triangles and solid line: after 180 m

Figure 3.26a–c shows the AFM images of the near-field etching.

Figure 3.27a shows the obtained values of surface roughness R_a, indicating the value decreased from 0.25 to 0.16 nm after 60 min etching. This was similar decrease as shown in the flat substrate. In addition, from Fig. 3.27b, the $R(l)$ value decreased especially less than 100-nm scale size (l).

As of the sample of the 3D structure, we performed the near-field etching on diamond mesa structure [27]. To fabricate the p-n junction of the diamond, the lateral-type structure developed. First, the diamond mesa structure of p-type is fabricated. When the mesa structure is fabricated along [110] orientation, the n-type diamond is selectively deposited at the side of the p-type mesa structure with $[1\bar{1}1]$ direction [28].

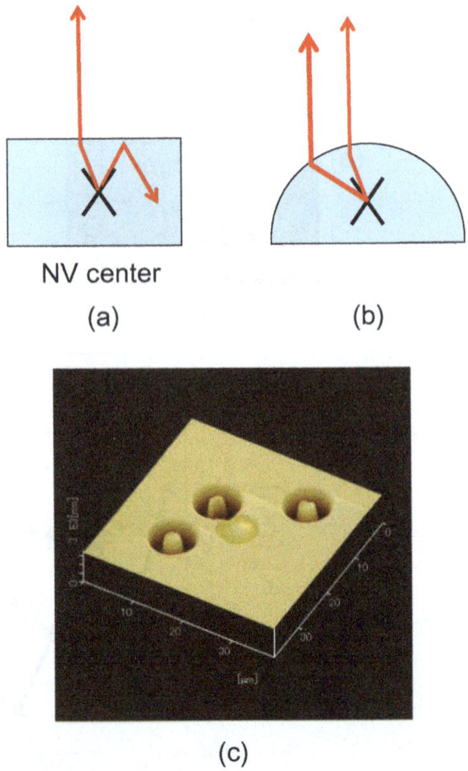

Fig. 3.23 **a** Due to the high refractive index of diamond, the critical angle becomes small. **b** By introducing the SIL structure, stronger light output can be obtained. **c** AFM image of the fabricated SIL structure

Figure 3.28a shows the AFM image of the prepared diamond mesa structure using the dry etching, indicating the side wall of the mesa structure has a corrugation. The corrugation might be transferred from that of the resist of the mask in through the dry etching process. To make it smooth, we performed the near-field etching using 532 nm light and oxygen in air. Figure 3.28b shows the AFM image after the near-field etching, in which etching time was 30 min. In the experiment, the incident light had a linear polarization along $y = x$.

From the cross-sectional profiles along the edges of the mesa structure shown in Fig. 3.28c, d, the near-field etching resulted in the realization of fine edge structures. In addition, by comparing the values of the surface roughness R_a on both the top and bottom of the mesa, the drastically decrease in the R_a value after the near-field etching (see Fig. 3.28e).

We also checked the polarization dependence of the near-field etching. Figure 3.29a shows the SEM images of the diamond mesa structure. From the image, the sidewall has large corrugations as much as wavelength size. In order to decrease these corrugations, we have done the near-field etching by controlling the polarization. We compared two incident light polarization conditions. To investigate the polarization dependence with respect of the corrugations, we irradiated the light with an incident

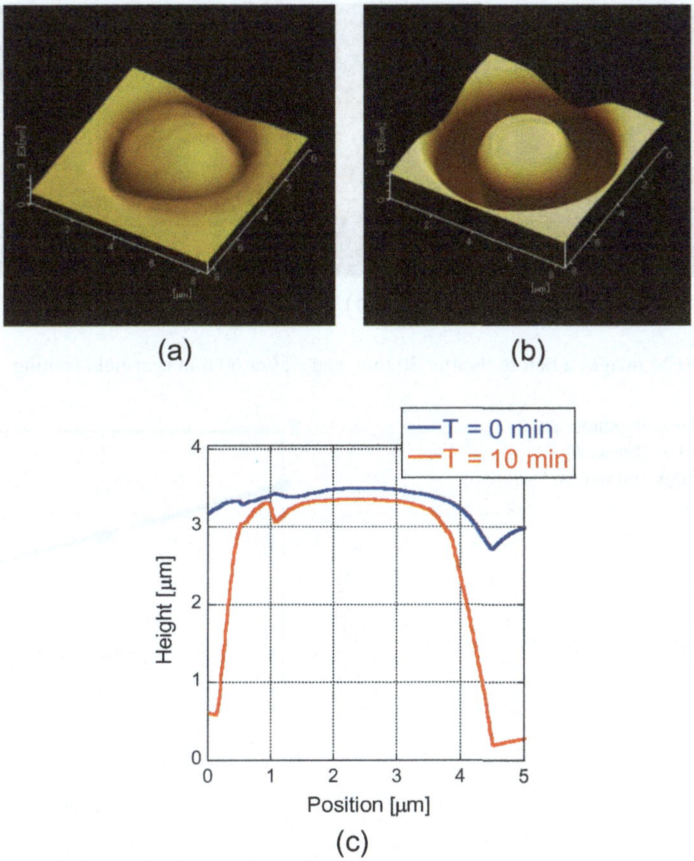

Fig. 3.24 AFM images of the **a** before and after the near-field etching. The etching time was 10 min. **c** Cross-sectional profiles along **a** and **b**

Fig. 3.25 Schematic of the lens array. *R*: curvature radius was 100 mm, *P*: pitch of the lens was 1.015 mm

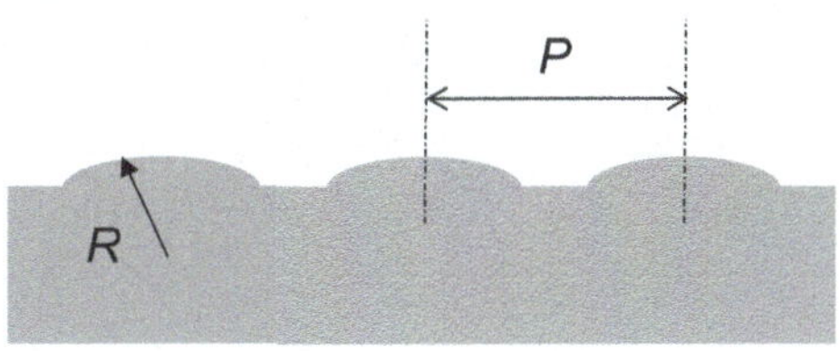

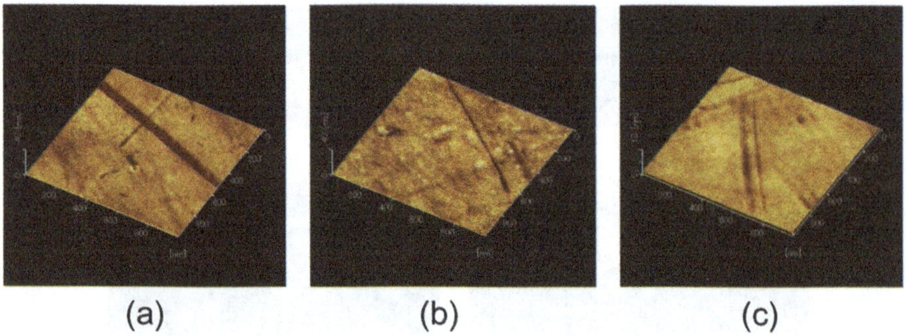

Fig. 3.26 AFM images **a** before, **b** after 30 min, and **c** after 60 min near-field etching

Fig. 3.27 Time dependence of **a** surface roughness R_a and **b** $R(l)$ (determined by (3.2))

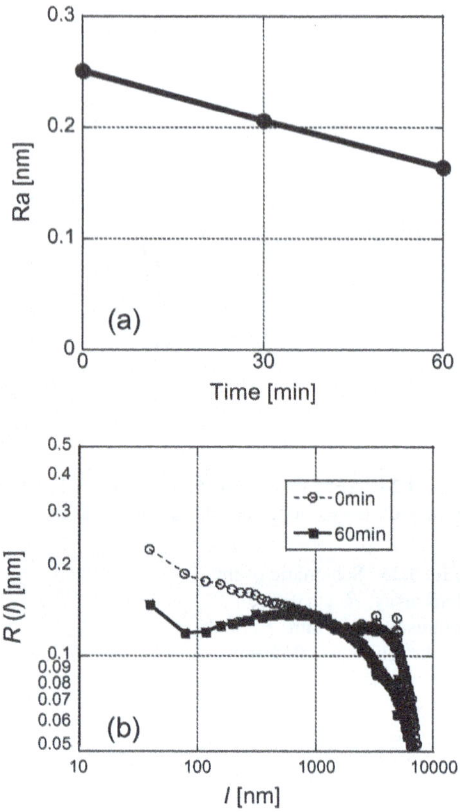

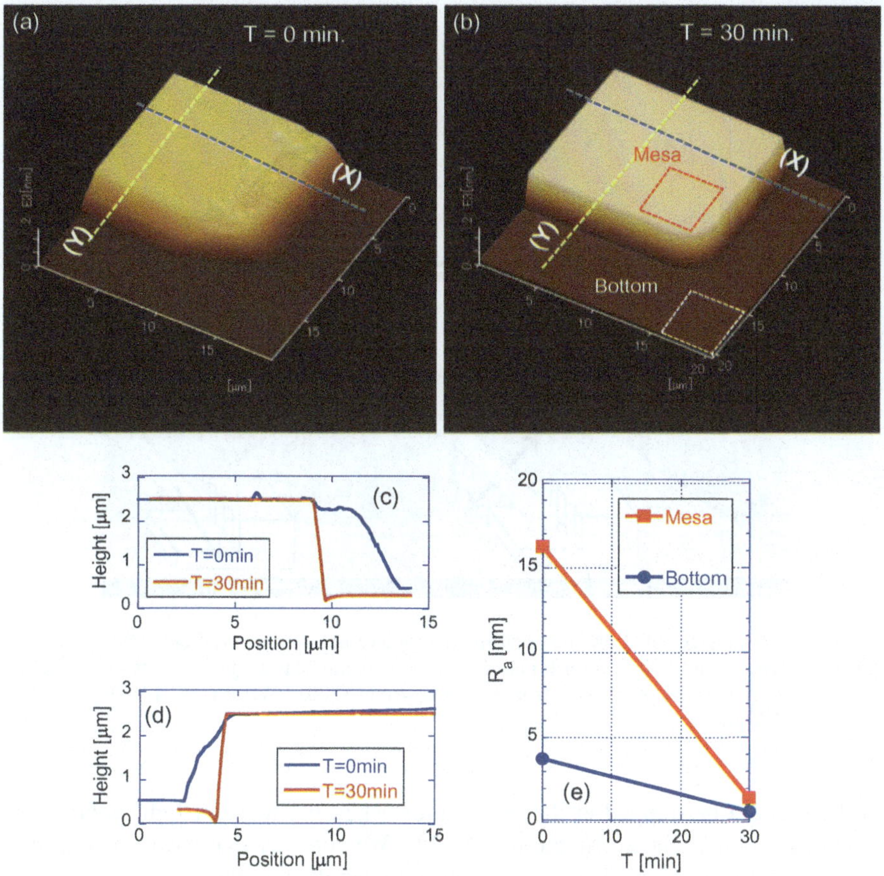

Fig. 3.28 AFM images **a** before and **b** after 30 min near-field etching. Cross-sectional profiles along **c** (X) and **d** (Y) in **a** and **b**. **e** Time dependence of surface roughness R_a on top and bottom of the mesa. These figures are reprinted with permission from [27], Copyright © 2014 WILEY-VCH Verlag GmbH & Co. KGaA, Weinheim

angle of 45° with respect to the diamond substrate with polarization parallel along the corrugation (see the inset of Fig. 3.29b), and an incident angle of 45° with respect to the diamond substrate with polarization perpendicular to the corrugation (see the inset of Fig. 3.29c).

Figure 3.29b, c shows the SEM images of the diamond mesa sidewall after the near-field etching with the corresponding etching conditions in the respective inset. The etching times were 30 min. It is noted that the sidewall corrugation decreased with polarization perpendicular with the corrugation (Fig. 3.29c). When the irradiation was oblique with respect to the diamond substrate, we obtained large absorption cross section of the incident light. Furthermore, perpendicular polarization to the corrugation induces a stronger field enhancement, because the carriers excited by

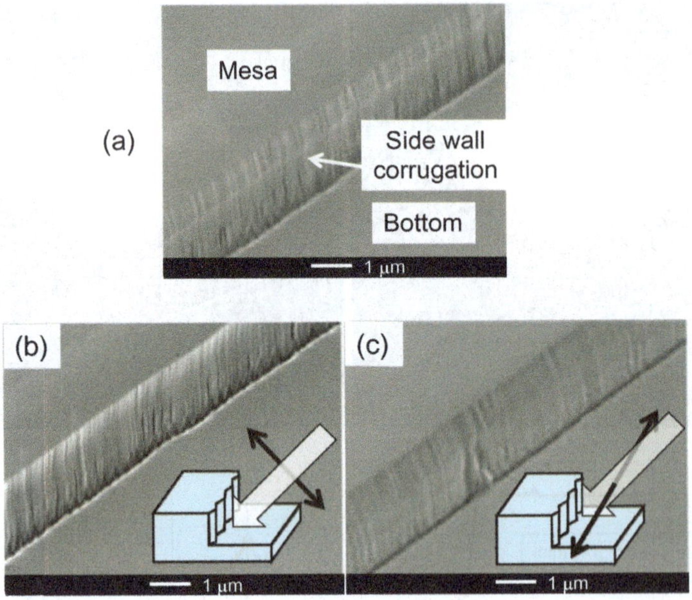

Fig. 3.29 SEM images of **a** the diamond mesa structure before the near-field etching, after the near-field etching with polarization **b** parallel and **c** perpendicular along the corrugation. These figures are reprinted with permission from [27], Copyright © 2014 WILEY-VCH Verlag GmbH & Co. KGaA, Weinheim

the light could be trapped at the corrugation when the incident light polarization is perpendicular to the corrugations [29–32]. We consider that such a strong field enhancement resulted in the enhancement of higher etching rate.

We deposited a diamond n-layer on the diamond mesa to check the effect of the near-field etching. As is described above, the mesa structure is fabricated along [110] orientation, the n-type diamond is selectively deposited at the side of the p-type mesa structure (see Fig. 3.30a) [28, 33]. We deposited a n-type diamond on the p-type diamond mesa of before the near-field etching (Fig. 3.30b) and of after the near-field etching (Fig. 3.30c). From the comparison of these images, we found a step-like structure on the sidewall without the near-field etching (Fig. 3.30b), while a flat structure was obtained at the sidewall (Fig. 3.30c).

From those results, the near-field etching resulted in improving the quality of the deposited materials after the near-field etching, as well as reducing the surface roughness.

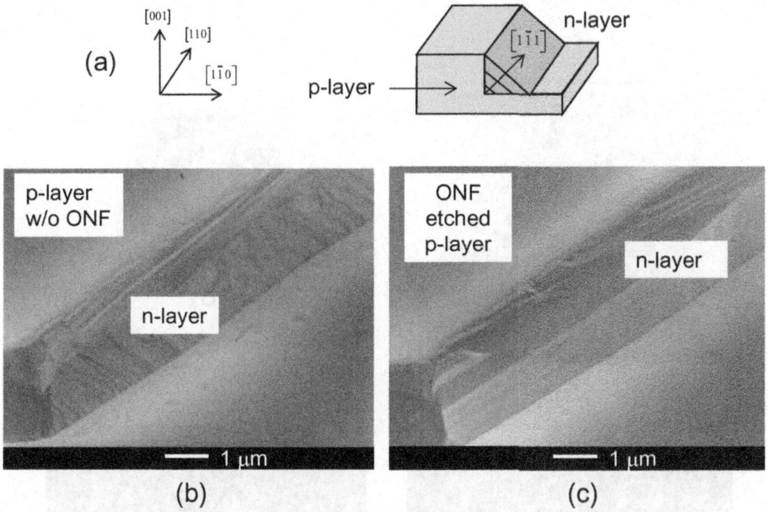

Fig. 3.30 **a** Schematic of the p-type diamond mesa structure. The n-type diamond selectively deposited at the side wall of the diamond mesa. **b** SEM image of the deposited n-layer of diamond on Fig. 3.29a (before the near-field etching. **c** SEM image of the deposited n-layer of diamond on Fig. 3.29c (after the near-field etching with perpendicular polarization). These figures are reprinted with permission from [27], Copyright © 2014 WILEY-VCH Verlag GmbH & Co. KGaA, Weinheim

3.6 Near-Field Makes Diamond More than Brilliant

We applied the near-field etching on the diamond jewelry. Ms. Ota (The University of Tokyo) kindly provided her own diamond pendant (see Fig. 3.31a) as a sample. After taking out the diamond from the base of the pendant, the diamond jewelry was irradiated using 325 nm light (see Fig. 3.31b). We used oxygen as an etching gas. By comparing the AFM images between before and after the near-field etching, the surface roughness R_a was decreased from 0.505 nm (Fig. 3.31c) to 0.375 nm Fig. 3.31d after 30 min.

Because the area of AFM image is limited in narrow area as small as several micron, we obtained the certification of the diamond to check the entire of the diamond. Figure 3.32a, b shows the heart and cubit patterns before and after the near-field etching, respectively. As indicated by the dashed circles, although the black bow tie patterns were not resolved before the near-field etching, we could see clear bow tie patterns after the near-field etching. In addition, the culet size was decreased from small (Fig. 3.32c) to none (Fig. 3.32c) after the near-field etching. From these results, the near-field etching resulted in the flattening entire diamond jewelry structure of round brilliant cut.

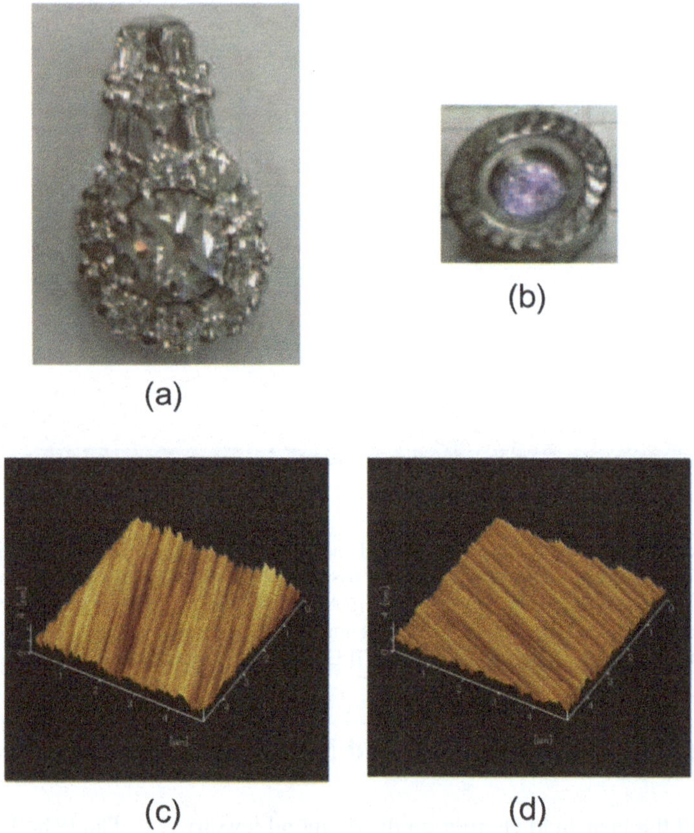

Fig. 3.31 a Photograph of the diamond pendant. b Photograph of the diamond pendant during the near-field etching. AFM images of the diamond pendant c before and d after the near-field etching. The surface roughness R_a was c 0.505 nm and d 0.375 nm

3.7 Power and Wavelength Dependence on the Near-Field Etching

In the process of the near-field etching, we examined the etching parameters. First of all, we checked the power dependence of the incident light [34]. For this purpose, we used nanodiamond as a sample, because from the change of the size, it could be evaluated the etching rate.

Figure 3.33 shows AFM images before and after the near-field etching with different incident power. We used 266 nm light (4th harmonic light of YAG laser) with 100 Hz and $\phi = 120$ mm. We used oxygen in the air as for the etching gas. To evaluate the etching rate, we obtained AFM images at the same positions between before and after the near-field etching. To eliminate the measurement errors, we obtained AFM images at three different positions. We irradiated laser with different power at each

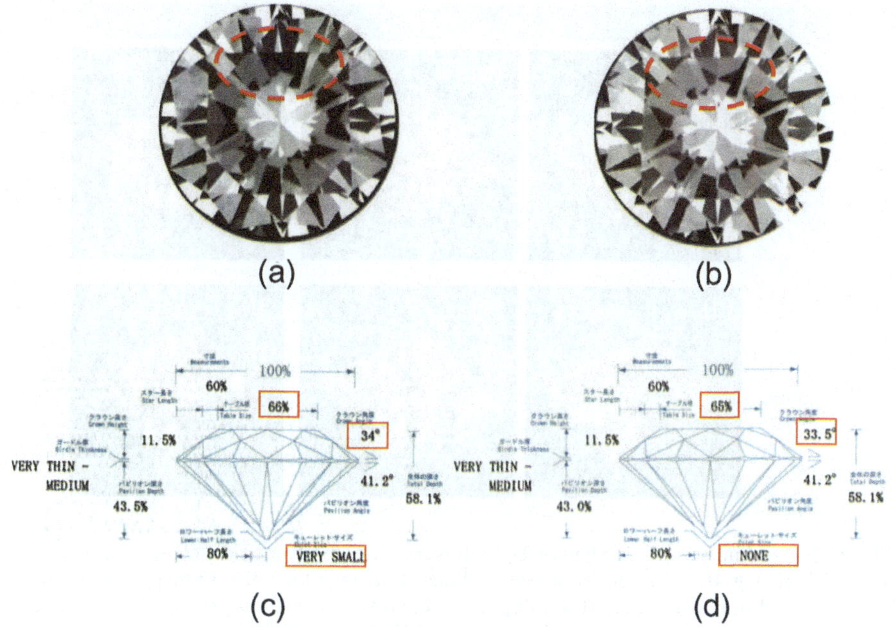

Fig. 3.32 Certification of the diamond. Heart and cubit patterns of **a** before and **b** after the near-field etching. The details of the diamond dimensions of **b** before and **c** after the near-field etching

position. Figure 3.33 shows the AFM images of the nanodiamond. We calculated the volumes at three different nanodiamonds at different AFM images indicated by circles in Fig. 3.33a–c.

Figure 3.34 shows the power dependence of the etching volume. From the results, the power dependence has linear dependence. If the photon absorption process is governed by the multi-photon process via dipole-forbidden state, or second harmonic generation, the power dependence should be fitted by the square. Therefore, this linear dependence might be caused by the excitation of dipole-forbidden state with a single photon and resultant dissociation process. Among the dipole-forbidden states in oxygen, the excited singlet state ($^1\Delta_g$) could be one of the possibility because it has high reactivity [35]. Although the transition from ground triplet ($^3\Sigma_g$) to $^1\Delta_g$ is spin forbidden transition, it can be excited by the generation of the ONF at the protrusions.

Next, we evaluated the wavelength dependence of the near-field etching [34]. For this purpose, we used TiO_2 line and space (LS) structures. The TiO_2 LS was prepared using the deep-ultraviolet (DUV) interferometric lithography method reported in the literature [36]. The structure was fabricated using photosensitive metal-oxo clusters, which are cross-linked and mineralized by DUV laser irradiation. Titanium propoxide ($Ti(OPr)_4$) was used as the metal alkoxide, and methacrylic acid (MAA) was used as the ligand to form a metal complex. Water and isopropanol were added to adjust the

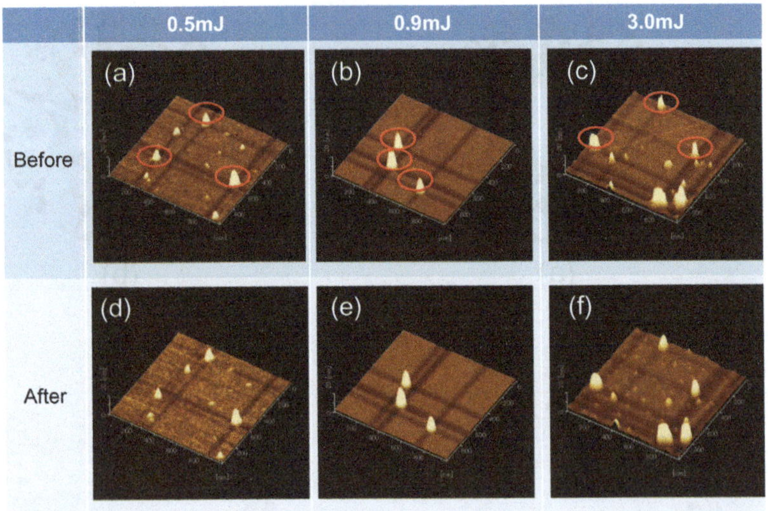

Fig. 3.33 AFM images of nanodiamond of before and after the near-field etching. **a–c** Before the near-field etching. **d–f** After the near-field etching. Irradiation time was 30 min. Same positions between **a** and **d**, **b** and **e**, **c** and **f**. Laser power were **d** 0.5 nJ, **e** 0.9 mJ, **f** 3.0 mJ. The figure is reprinted with permission from [34], Copyright © 2017 Springer Nature

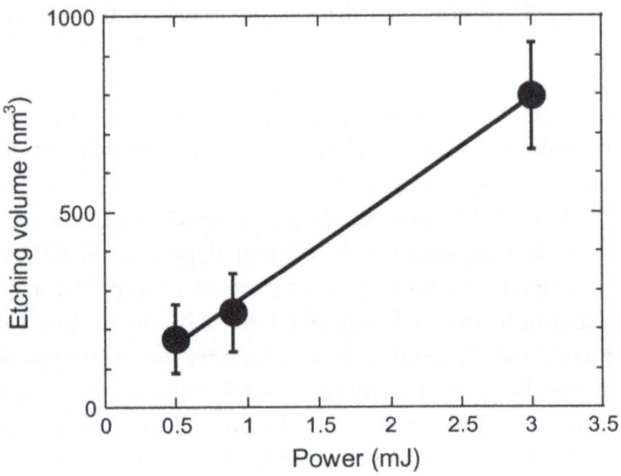

Fig. 3.34 Power dependence of the etching volume. The figure is reprinted with permission from [34], Copyright © 2017 Springer Nature

physicochemical properties of the resist. The DUV irradiation led to the cross-linked structure which resulted in the formation of TiO_2 after treatment of the non-exposed part with cyclohexanone.

We used oxygen gas in the air as an etching gas. To examine the wavelength dependence, we used light emitting diode (LED) with different wavelength 285, 300, 310, 365 nm. The light intensity of all LEDs was set to 5.0 mW/cm^2. Thus, the

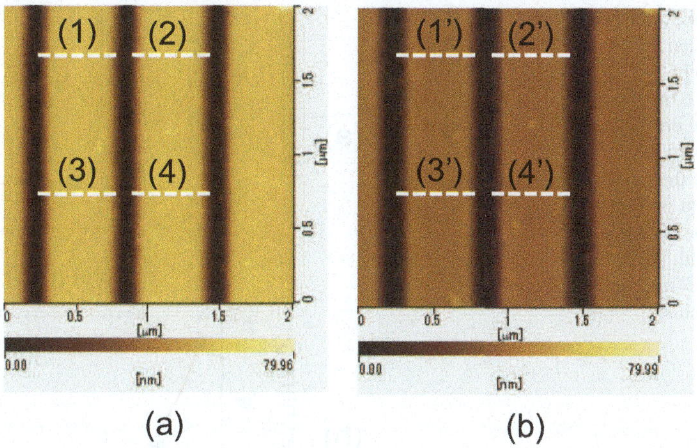

Fig. 3.35 AFM images of TiO_2 LS structures. **a** Before and **b** after the near-field etching. The height of the TiO_2 LS structures was taken at four different positions. (1')–(4') correspond the same position taken before the near-field etching at (1)–(4). The figures are reprinted with permission from [34], Copyright © 2017 Springer Nature

incident wavelength was longer than the dissociation wavelength of O_2 (242 nm) [9, 10].

To evaluate the height change of TiO_2, we used AFM with a *Sampling Intelligent Scan mode* (Hitachi High-Tech Science Corporation). Figure 3.35 shows the AFM images of TiO_2 LS structures. The value of the height was calculated from the average height obtained from the cross-sectional profiles. By comparing the height of LS structures at the same positions before and after the near-field etching, we evaluated the time dependence of the height change (see Fig. 3.36).

The height changes were fitted by the following single exponential curve,

$$h(t) = A_1 \exp(-t/t_1) + h_0 \tag{3.4}$$

where A_1 is the constant, t_1 is the time constant of the height decrease, h_0 is the initial height. The fitted curves are shown in Fig. 3.36a–c. From the fitted values, we plot the photon energy dependence of the value t_1 in Fig. 3.37.

This result show that t_1 decreases as the incident photon energy increases. To dissociate the oxygen, it is required higher photon energy than 6 eV to excite the excited state of $B^3\Sigma_u^1$. Therefore, to excite $B^3\Sigma_u^1$ via the intermediate state or by the second harmonic generation, a value more than half of 6 eV (3 eV) is required. However, in the case of the result with 3.40 eV (365 nm), the etching height unchanged. From the wavelength dependence observed here, it is required wavelength less than 365 nm for the near-field etching using LED light source.

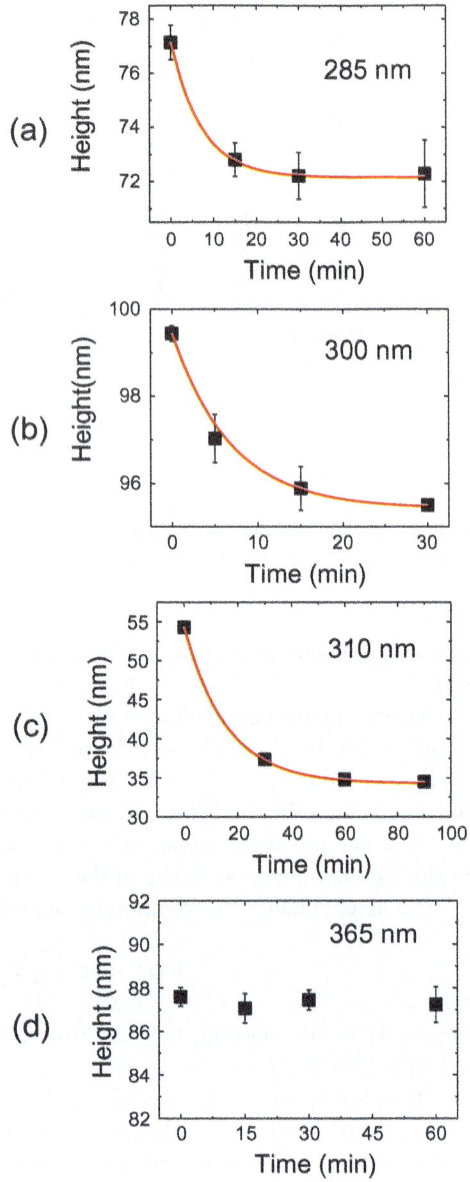

Fig. 3.36 Wavelength and time dependence of the etching height. **a** 285 nm, **b** 300 nm, **c** 310 nm, and **d** 365 nm. The error bar corresponds to the standard deviation of the height for four different positions (1)–(4). The figure is reprinted with permission from [34], Copyright © 2017 Springer Nature

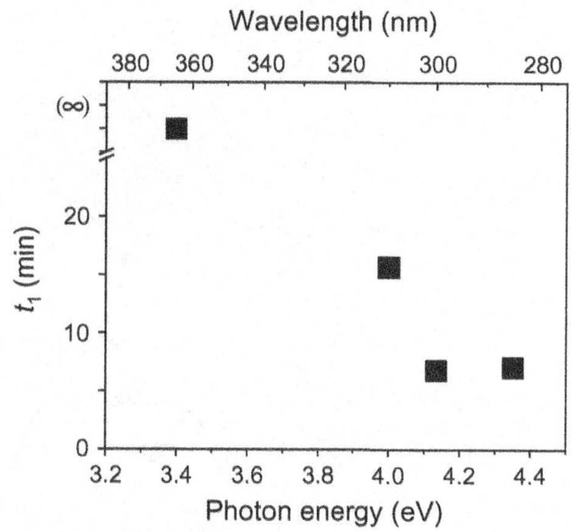

Fig. 3.37 Wavelength dependence of the etching height. The figure is reprinted with permission from [34], Copyright © 2017 Springer Nature

3.8 Improving the Properties of Materials

As described in Sect. 3.5, it is expected that the diamond NV could be a good magnetic sensor with sub pT sensitivity (see Fig. 3.38a) [37]. In addition, by introducing NV in nanoscale diamond (nanodiamond) [38] can be utilized to visualize the magnetic field distributions in the living cells [39, 40]. However, the nanodiamond has some non-radiative recombination centers, because the nanodiamonds are obtained through milling of microdiamond powder [41]. Therefore, the excited carries in the nanodiamond can be trapped, which resulted in the decrease in the emission intensity. To realize the etching away the non-radiative recombination centers, we performed the near-field etching [42].

Figure 3.39a, b shows AFM images before and after the near-field etching of nanodiamond, respectively. The near-field etching was performed using He-Cd laser with a wavelength of 325 nm light and oxygen in the air. To check the etching properties, we obtained the cross-sectional area from the AFM images. p and s show the parallel and the perpendicular with respective to the incident light polarization (show as E in Fig. 3.39a). Figure 3.39c shows the time dependence of the cross-sectional area. From the results, the etching rate along p-direction was larger than that of s-direction. This might be caused by the carriers concentration at the edge of the material [27, 29, 30, 32].

To check the optical properties of the nanodiamond after the near-field etching, we compared the cathodoluminescence (CL) [42]. Each nanodiamond had 500 NV. Because the size of the nanodiamond decreased as shown in Fig. 3.39, we obtained the normalized CL intensity. As shown in Fig. 3.40, the CL intensity increased by 13.2 times after 30 min near-field etching. This result supports the postulate that the near-field etching could etch away non-radiative recombination centers at the surface of the nanodiamond.

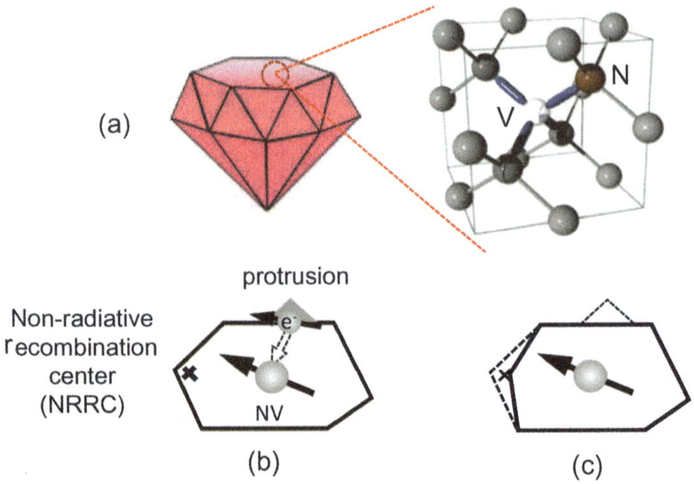

Fig. 3.38 Schematic of the diamond NV. N: nitrogen, V: vacancy. Schematic of nanodiamond **b** before and **c** after the near-field etching. Before the near-field etching, there are some non-radiative recombination centers and protrusions near the surface of the nanodiamond, which affect the spin property of NV. These figures are reprinted from [2] under the terms of the creative commons attribution license

To evaluate the spin properties of diamond NV, we obtained the optically detected magnetic resonance (ODMR) spectra [22]. In the electronic structures of diamond NV, the ground state has spin projections m_s of 0 and ±1. If the magnitude of magnetic field is zero, the m_s of ±1 are shrinked. The separation between $m_s = 0$ and ±1 is 2.87 GHz (see Fig. 3.41). When we excite the diamond NV with 532-nm light, the NV emit the light at 637 nm of zero phonon line. During the irradiation of 532-nm light, if we introduce the microwave at 2.87 GHz, resonant to the separation between $m_s = 0$ and ±1, spin projections is excited from $m_s 0$ to ±1. Then if we excite the diamond NV with 532-nm light, carriers are excited to the $m_s = \pm1$ of excited states. 30 % of carriers in $m_s = \pm1$ of excited states transfer to the non-radiative state of singlet $^1A_1$. Thus, the light intensity at 637 nm decrease (see Fig. 3.41). As a result, if we scan the frequency of magnetic wave, the spectrum of the emission intensity (ODMR spectrum) has a dip at the resonant frequency at 2.87 GHz (see Fig. 3.42a).

We performed the near-field etching using He-Cd laser with a wavelength of 532 nm and oxygen in the air. Figure 3.42a shows the typical ODMR spectra of before and after the near-field etching. By comparing the spectra, we found the near-field etching resulted in the decrease of the full width at half maximum (FWHM) of the ODMR spectra. Figure 3.42b shows the time dependence of FWHM values of the ODMR spectra as a function of the near-field etching time. From the results, the value of FWHM decreased by 14% by the near-field etching.

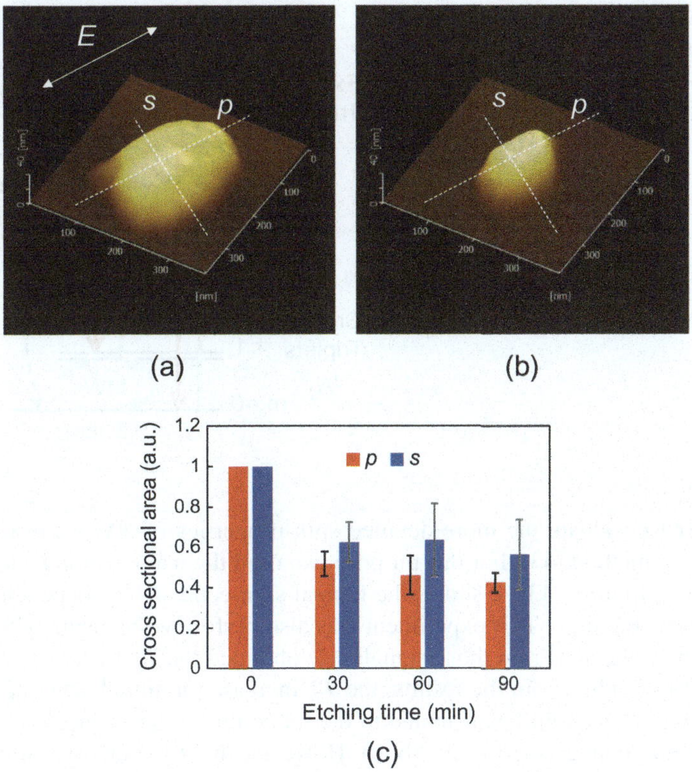

Fig. 3.39 AFM images of **a** before and **b** after the near-field etching. E represents the direction of the incident light polarization. p and s represent the direction parallel and perpendicular with respective to E. **c** Time dependence of the cross-sectional area along p and s. These figures are reprinted from [2] under the terms of the creative commons attribution license

Fig. 3.40 Time dependence of normalized CL intensity as a function of the near-field etching. These figures are reprinted with permission from [42], Copyright © 2015 Springer Nature

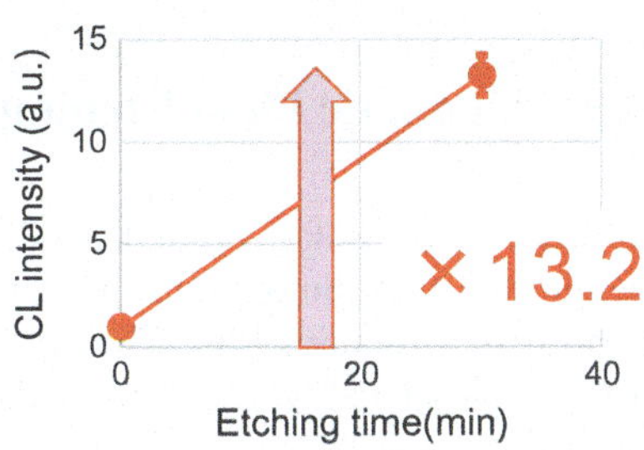

Fig. 3.41 Energy diagram of the diamond NV

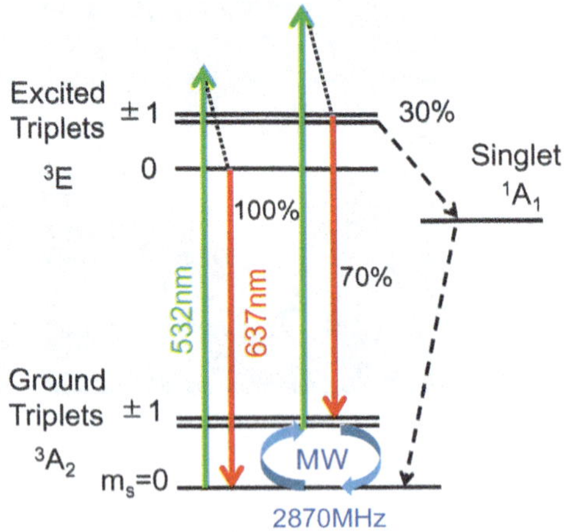

In order to evaluate the more detailed spin properties of NV, we evaluated T_2^* by spin-echo method [43]. For this purpose, we used the nanodiamond with the NV number of 1. Figure 3.43a–c shows the typical spin-echo signals depending on the near-field etching time. This experiment was also performed by using 325 nm light and oxygen in the air. From those signals, we obtained the T_2^* by fitting the data as shown in Fig. 3.43d. From the results, the T_2^* increased at initial stage of the near-field etching. This might be attributed to the reduction of the surface protrusion by the near-field etching (see Fig. 3.38b, c). However, the T_2^* decreased after 60 min of the near-field etching. The reason why the decrease of the T_2^* is not unclear; however, it might be caused by oxygen molecule's adsorption onto the diamond surface during the near-field etching process, which can affect the NV-center spin due to their paramagnetic nature [44].

3.9 Magnetic Field Etching

To evaluate the polarization dependence on the near-field etching, we compared the height and morphology changes of the LS structures [45]. We used ZrO_2 as the nanostripe structures [46]. The fabrication process of ZrO_2 LS is similar to that of TiO_2 as shown in Fig. 3.35 [36]. In the preparation process, we used $Zr(OPr)_4$ instead of $Ti(OPr)_4$ for TiO_2 LS.

We used He-Cd laser with a wavelength of 325 nm (excitation power: 0.8 W/cm^2) and oxygen in the air as a gas source. Thus, the wavelength is longer than the dissociation wavelength of oxygen [9, 10]. The absorption edge wavelength of a ZrO_2 is around 250 nm [36]. Therefore, the effect of the carrier generation in the ZrO_2 LS patterns is excluded. To evaluate the changes of ZrO_2 LS structure, we used AFM with a *Sampling Intelligent Scan* mode (Hitachi High-Tech Science Corporation). Images (a) and (b), and (c) and (d) were obtained at the same respective positions.

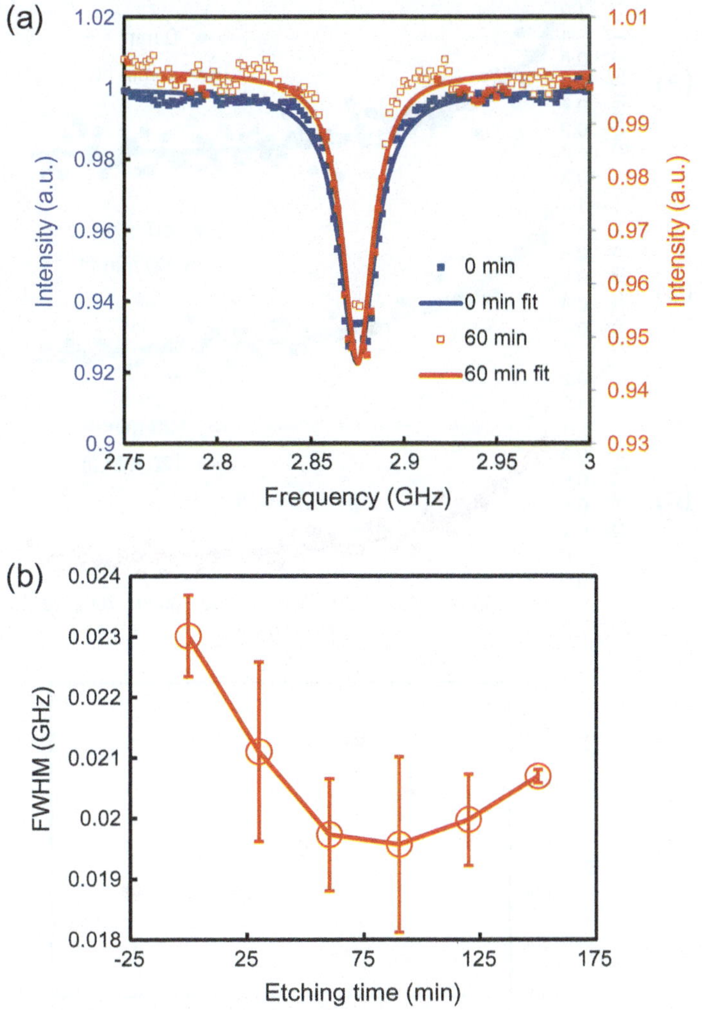

Fig. 3.42 **a** Time dependence of ODMR spectra. **b** Time dependence of FWHM of ODMR spectra as a function of the near-field etching. These figures are reprinted from [2] under the terms of the creative commons attribution license

Figure 3.44a, c shows the AFM images before etching of ZrO_2. From the AFM images, the LS had a period of 598.4 ± 14.7 nm and a width (w_b), of 438.1 ± 13.0 nm. In addition, AFM images were taken at the same positions before and after 3-h irradiation, respectively (see Fig. 3.44b, d).

To evaluate the shape changes, we obtained the cross-sectional profiles along the x-direction in the AFM images (see Fig. 3.45a–d). Figure 3.45a–d show cross-sectional profiles by the x- and y-polarizations, respectively. The dashed and solid curves in the upper position in each figures correspond to they were taken before and after

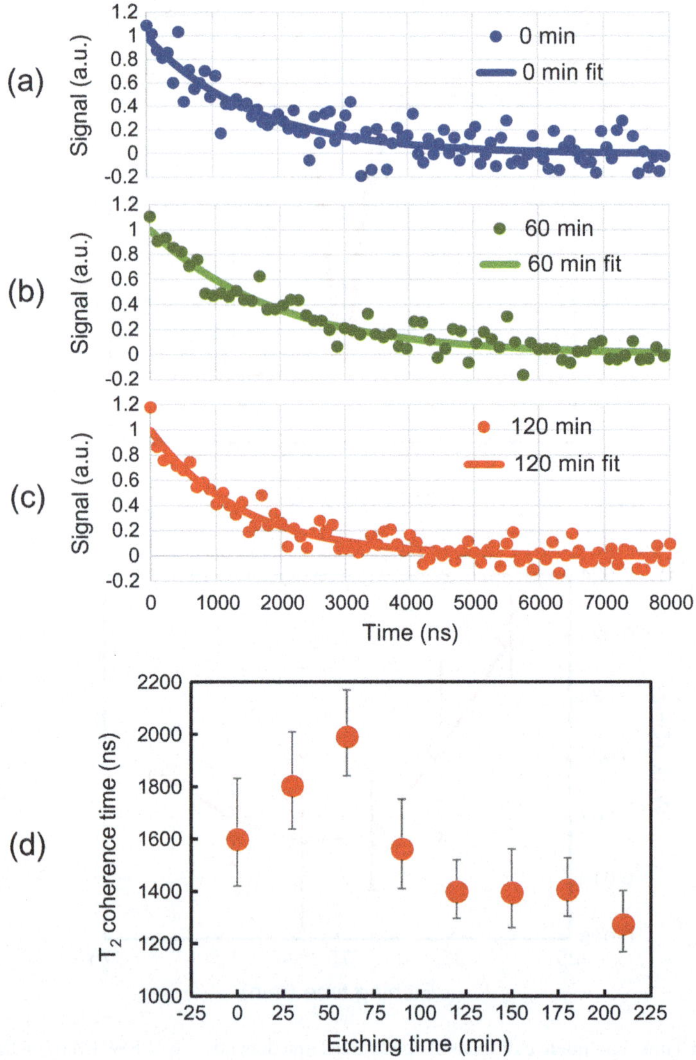

Fig. 3.43 Normalized spin-echo signals **a** before, **b** after 60 min, and **c** after 120 min near-field etching. **d** Obtained T_2^* time. These figures are reprinted from [2] under the terms of the creative commons attribution license

etching, respectively, at the same y-position. The solid curve in the lower position in each figures in Fig. 3.45 shows the cross-sectional profiles of height differences in the before and after the near-field etching. From these profiles, it was found that the near-field etching with x-polarization resulted in uniform etching through the land area; however, with y-polarization resulted in non-uniform etching. To support this postulate, we evaluated that the surface roughness R_a, at the land area. The value

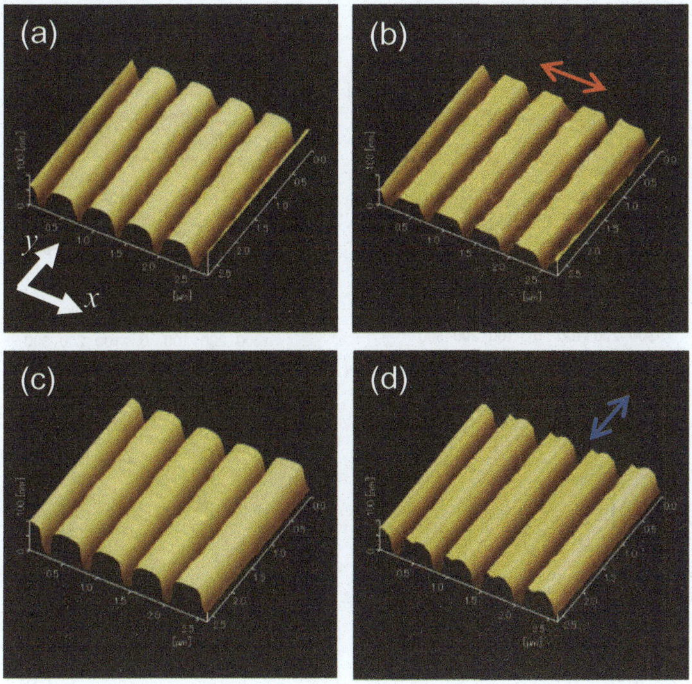

Fig. 3.44 AFM images of ZrO_2 LS structures of before ((**a**) and (**c**)) and after ((**b**) and (**d**)) near-field etching. The etching time was 3-h. Polarization: **b** Perpendicular (x-polarization) and **d** parallel (y-polarization) to the stripes. The figure is reprinted from [45] under the terms of the creative commons attribution license

of R_a decreased from 1.14 to 0.60 nm with the x-polarization etching, while that increased from 1.25 to 3.88 nm in the y-polarization etching.

To support the polarization dependence of etching property, we evaluated the electric and magnetic field distributions numerically. To calculate the fields on the LS structure, we used a finite-difference time-domain (FDTD) method (Fujitsu Poynting for Optics). In the simulation, the parameters were following:

1. The cell size was 2 nm (x) $\times$ 1 nm (y) $\times$ 2 nm (z).
2. The model size was 600 (x) $\times$ 1 (y) $\times$ 2000 (z) cells.
3. Periodic boundary conditions (the x- and y-axes) and a perfectly matched layer (PML) absorbing boundary condition (the z-axis).
4. A plane wave with an homogenous intensity distribution in the xy-plane was set at 600 nm from the top surface of the ZrO_2 structure.
5. The refractive index of ZrO_2 was 1.8 [36].
6. The peak intensity of the incident electric field was set to 1 [V/m].
7. A periodic structure consists of land (50-nm height, w-nm width) and space (100-nm spacing).

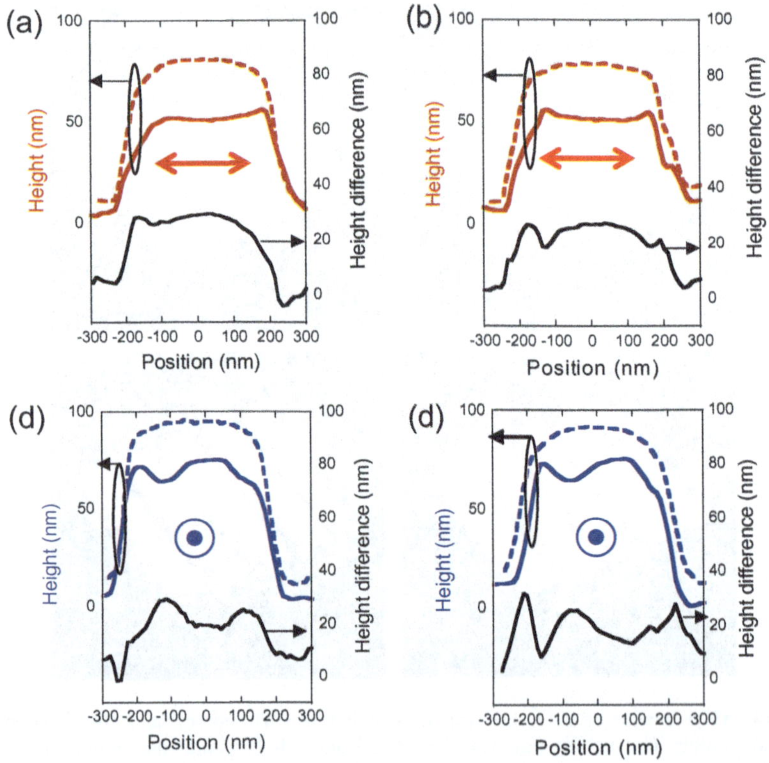

Fig. 3.45 Cross-sectional profiles of AFM images (left axis) of before (dashed line) and after (solid line) the near-field etching. **a** and **b** were taken with x-polarization. **c** and **d** were taken with y-polarization. Solid line at the bottom of each figures (right axis) were the height difference between before and after the etching. The figure is reprinted from [45] under the terms of the creative commons attribution license

Figure 3.46a–d shows the polarization dependence of the electric field intensity $|E^2|$ and magnetic field intensity $|H^2|$ distributions, in which Fig. 3.46a, c for the perpendicular (x) polarization and Fig. 3.46b, d for the parallel (y) polarization. From these results, it was found that $|E^2|$ was enhanced at the corner of the land for both x- and y-polarizations. However, the $|H^2|$ was different dependent on the polarization, in which $|H^2|$ was focused at the center of the land structure for x-polarization (Fig. 3.46c) and $|H^2|$ has two peaks near the corner (Fig. 3.46d).

Based on the FDTD calculations for different land width w, we compared the calculated results of magnetic field and obtained results of the height change profile. Figure 3.47 shows the comparison of them for narrow ($w = 410$ nm) and wide ($w = 450$ nm) LS structures.

The electric field E is localized at the corner of the land structure, which is originated from the edge effect [29]. Therefore, the calculated E distributions do not explain the polarization dependence of the height changes. In contrast, the magnetic

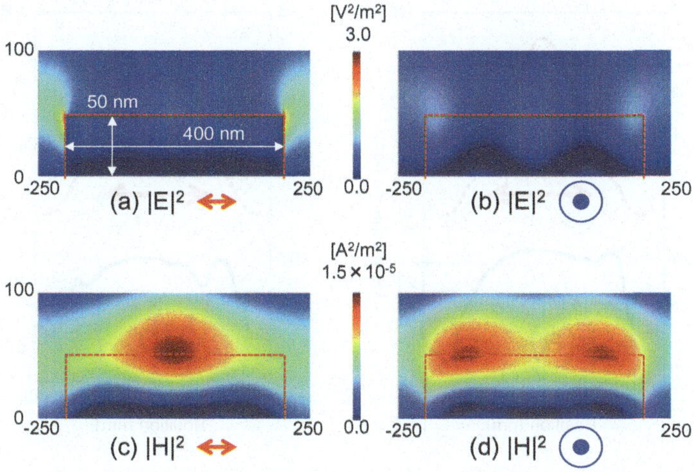

Fig. 3.46 Calculated FDTD results of polarization dependence for land width w of 400 nm. Electric field intensity $|E|^2$ with **a** perpendicular polarization to the LS (x-polarization) and **b** parallel polarization to the LS (y-polarization). Magnetic field intensity $|H|^2$ with **c** perpendicular polarization to the LS (x-polarization) and (**d**) parallel polarization to the LS (y-polarization). The figure is reprinted from [45] under the terms of the creative commons attribution license

field H is focused at the center of the land structure. In addition, focused profiles of the H depends on the w. As shown in Fig. 3.47, the height change profiles are in good agreement with the H distributions. The height change might be governed by the near-field distributions, and it could be reasonable to be determined by the H distributions, not the E distributions.

3.10 Near-Field Assisted Desorption

To realize ultra-flat surface, we developed near-field assisted desorption process using transparent ceramics of alumina (Al_2O_3) [47]. Recently, alumina is expected to be used as a low-loss gain medium for the ceramic lasers [48, 49]. Because alumina has high breakdown threshold, the ceramic lasers could have high output power so that it realize laser-driven spark plugs for ignition systems in automobile engines [50, 51]. However, alumina is a hard material, and it is difficulty to realize flat surface by the mechanical polishing which leave mechanical scratches. Therefore, we applied near-field assisted desorption which reduce the surface roughness. We consider that the introduction of light irradiation during sputtering with Al_2O_3 nanoparticles on the alumina substrate will reduce the surface roughness by the near-field desorption. We used radio frequency (RF) sputtering Al_2O_3 nanoparticles deposition on an alumina substrate. The migration length of the nanoparticles on the surface depends on the Schwoebel barrier [52, 53] in the free energy profile by sputtering process.

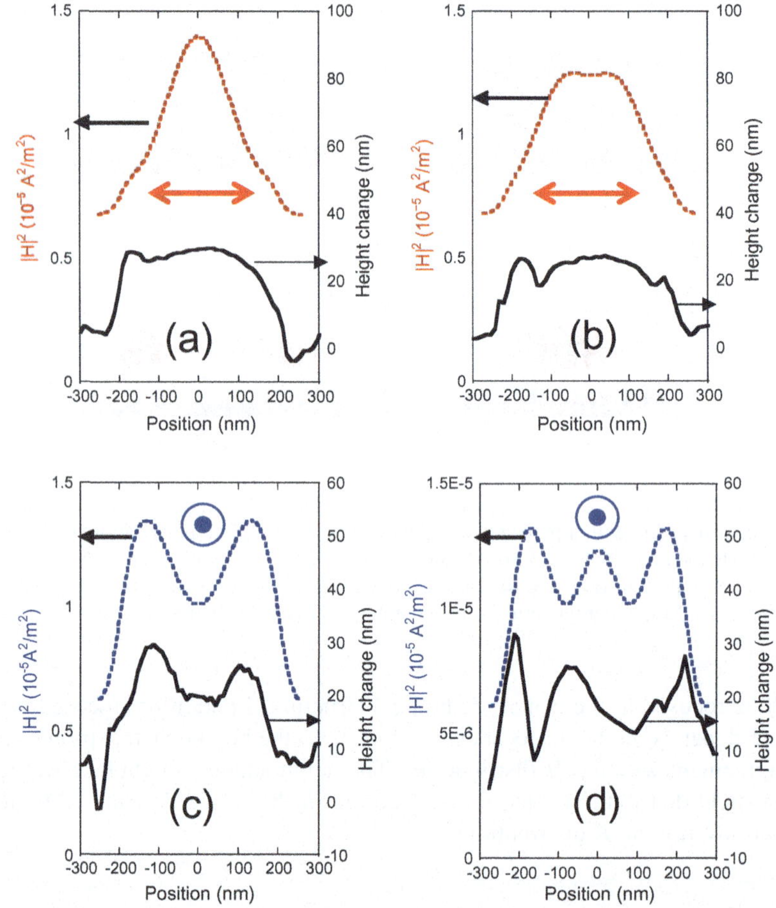

Fig. 3.47 Comparison between calculated results of magnetic field (dashed line and left vertical axis) and the etching height change (solid line and right vertical axis). **a** and **b** x-polarization. The calculated w and the initial widths, w_b, were **a** $w = 410$ and $w_b = 411.17$ and **b** $w = 450$ and $w_b = 454.45$ nm. **c** and **d** y-polarization. **c** and **d** y-polarization. **c** $w = 410$ and $w_b = 414.06$ and **d** $w = 450$ and $w_b = 445.11$ nm. Red and blue solid curves correspond to the calculated magnetic field intensity for x- and y-polarizations, respectively. Black solid curves correspond to the height change obtained from Fig. 3.45. The figure is reprinted from [45] under the terms of the creative commons attribution license

The scratches on the surface result in increase in the Schwoebel barrier height, and thus, the migration length becomes short when the nanoparticles comes to around the scratches. Therefore, the deposition rate of the Al_2O_3 nanoparticles increases at ridge of the scratches in comparison with that in flat surface. Hence, we could not repair the scratches by conventional deposition techniques. To realize the flat surface of substrate with the scratches, we deposited Al_2O_3 with irradiation of visible light. To realize near-field desorption, we selected wavelength to be transparent for alumina

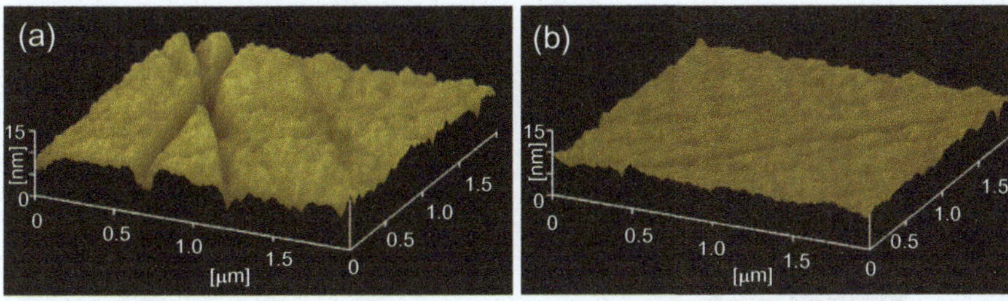

Fig. 3.48 AFM images of the alumina substrate surface. **a** After RF sputtering without irradiation. **b** After RF sputtering with visible light irradiation with $\lambda = 473$ nm, the sputtering time was 30 min. The figure is reprinted from [47] under the terms of the creative commons attribution noncommercial license

substrate. Because the absorption band edge wavelength of Al_2O_3, λ_{ab}, is 260 nm [54], we irradiated the light with $\lambda = 473$ nm, and the incident power density was 2.7 W cm^2. The power density was low enough to prevent the substrate heating. In this process, the ridges of the scratches work as the ONF source. Then, the generated ONF increases in migration length of the Al_2O_3 nanoparticles, resultant desorption from the ridge area [31, 55–57]. However, the Al_2O_3 nanoparticles on the flat surfaces where no ONF generates have the same deposition rate. Based on the above discussions, we expect deposition at the ridges was suppressed by the near-field assisted desorption. Finally, the scratches are expected to be repaired. Figure 3.48a, b represents AFM images before and after (30 min sputtering time) the near-field assisted sputtering of Al_2O_3. We found that the average scratch widths decreased from 128 (Fig. 3.48a) to 92 nm (Fig. 3.48b) by the near-field desorption, in which the average scratch widths were obtained by statistical analysis using a Hough transform of AFM images. In addition, we found that the average depth decreased from 3 to 1 nm.

To confirm the near-field assisted selective desorption of nanoparticles at the nanoscale structure, we deposited Al_2O_3 nanoparticles on a sapphire substrate, because we could obtain a uniform atomic step structures on the sapphire substrate by thermal annealing [58] (see details in Sect. 3.11). Figure 3.49a, b shows respective AFM images taken after sputtering (sputtering time of 30 s), without and with visible light ($\lambda = 473$ nm) irradiation and used to the flattened grayscale images to clarify the deposition positions on the surface in Fig. 3.49c, d, which is obtained from Fig. 3.49a, b, respectively. Figure 3.49e, f shows the cross-sectional profiles along the dashed lines in Fig. 3.49a, c, respectively. From these images, we found that the conventional RF sputtering resulted in the selective deposition at the terrace edge. In addition, we also found that the visible light irradiation during RF sputtering resulted in no growth of Al_2O_3 nanoparticles at the terrace edges. These results confirmed the evidence of the near-field assisted desorption.

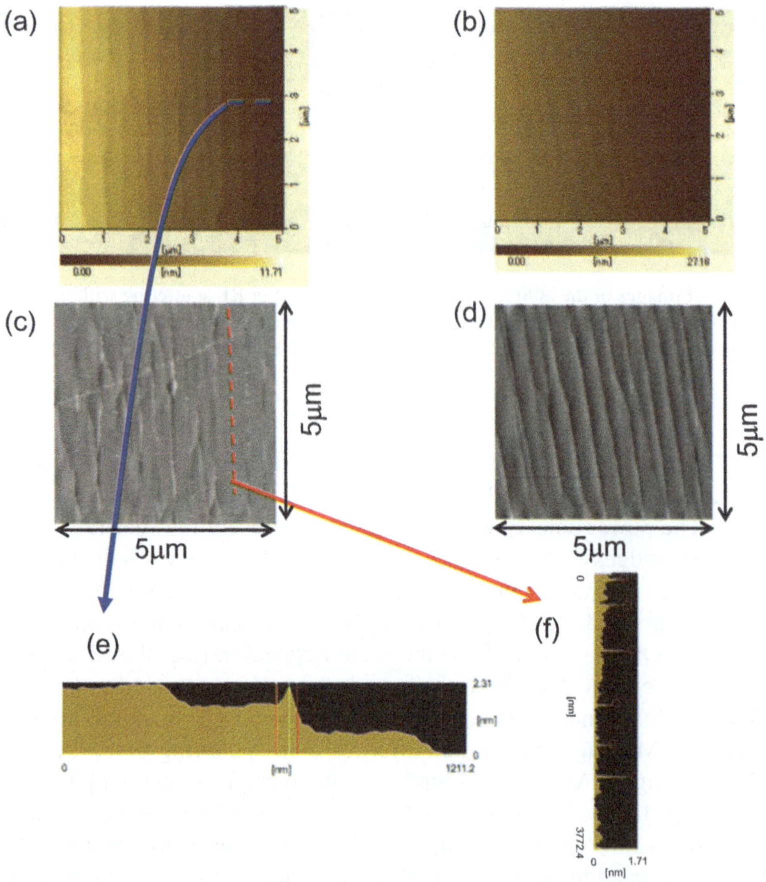

Fig. 3.49 AFM images of the sapphire substrate. **a** After RF sputtering without irradiation (conventional sputtering). **b** After visible light irradiation (near-field assisted sputtering). **c** and **d** represent the respective flattened AFM images of **a** and **b** in grayscale. **e** and **f** represent cross-sectional profiles along the dashed lines in **a** and **c**, respectively. The figure is reprinted from [8] under the terms of the creative commons attribution license

Based on the results of initial growth, we deposited alumina film using the near-field assisted sputtering. Figure 3.50 shows the comparison of time dependence of the surface roughness R_a between the conventional sputtering (without irradiation) and the near-field assisted sputtering (with visible light irradiation). As can be seen, R_a increased as the sputtering time increased by conventional sputtering (solid squares in Fig. 3.50). However, we obtained an ultra-flat alumina surface with $R_a = 0.08$ nm by near-field assisted sputtering after 600 s (solid circles in Fig. 3.50).

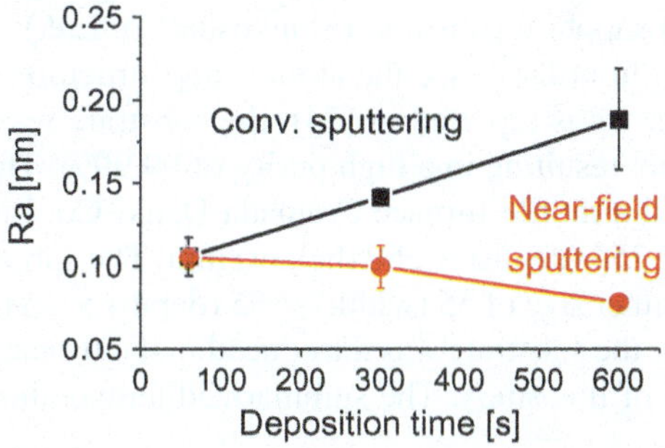

Fig. 3.50 Dependence of R_a on sputtering time. Solid black squares denote results for conventional RF sputtering, and solid red circles, for RF sputtering under visible light irradiation. The figure is reprinted from [8] under the terms of the creative commons attribution license

3.11 Related Techniques

Good-quality single-crystalline sapphire is required for preparation of various semiconductors because it has thermal stability. In order to grow a high-quality semiconductor film, an ultra-flat surface is required, to avoid to form dislocations in the grown film. Yoshimoto et al. had discovered the annealing conditions for an atomically flattened sapphire substrate at temperatures exceeding 1,000 °C [58]. The use of the sapphire substrate with (0001) crystalline plane results in an atomically flat surface. The ultra-flat surface of sapphire has a multiple of the atomic step height, in which the unit atomic step is $h_u = 0.21$ nm [58]. The atomic step height increases as the annealing temperature T_{ann} increases. It was found that there is threshold of T_{ann} to form multi-step formation [59];

- mono-step formation at $T_{\mathrm{ann}} < 1{,}200\,°\mathrm{C}$
- multi-step formation at $T_{\mathrm{ann}} > 1{,}200\,°\mathrm{C}$

Annealing time was investigated, revealing that over annealing time deform the atomic step [60]. Although we obtain the atomically flat sapphire by simple annealing process, the atomically flat area with an uniform planes of $(1\bar{1}00)$ is limited in 10 μm × 10 μm [61]. By introducing the atomically flattened sapphire substrate, there have been reported of improving the grown materials such as ZnO [62–64], GaN [65–67], diamond [68], and graphene [69]. In addition, the step structures were also available as a template for carbon nanotubes [70]. For future applications, we need to control uniformity of the atomic step structure over the substrate (mm range).

In this section, we introduced the "tempering" process [71] during thermal annealing of sapphire. This idea was given by Ms. Ota (The University of Tokyo), who got this idea during producing chocolate. To produce good taste chocolate, the tempering process is required [72]. At the beginning of chocolate production, cacao butter is heated to melt. Then, the temperature decreases to obtain solid state. During the cooling process, the temperature is maintained at some time and then increased again. This process results in single-phase cacao, and it is called as "tempering" [73].

We used a (0001) sapphire substrate (Shinkosha Co. Ltd.), which had 3 × 5 × 0.43 mm (thickness). In order to see the atomic step structure, we used a substrate with a 0.2° off angle. As is reported in [74], the substrate was prepared using the Kyropoulos technique, resulting in a high purity of 99.99%. The annealing process in this study, we used a muffle furnace (Yamada Denki Co. Ltd.), having an inner size of 200 (width) × 255 (depth) × 260 (height) mm. The substrate was placed on a ceramic pot with an inner size of 55 (width) × 55 (depth) × 35 (height) mm, and the ceramic pot including the substrate was introduced in the furnace. Figure 3.51 shows the temperature data of this study. The summarized temperature data are shown in Table 3.1.

Figure 3.52a–e shows typical AFM images of samples at the center of the substrate prepared using annealing process I (simple annealing at low temperature), process II (simple annealing at high temperature), process II' (simple annealing at high temperature with long time), process III' (multi-step annealing), and process III (tempering), respectively. Figure 3.52f is a wide-area AMF image of Fig. 3.52e

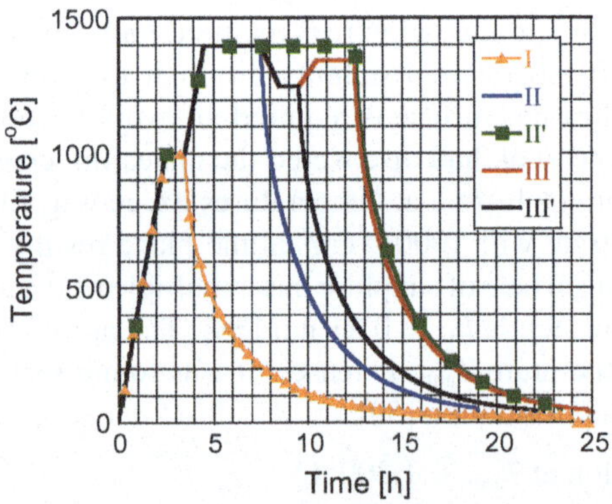

Fig. 3.51 Annealing temperature and time. Process I: 1,000 °C for 1 h. Process II: 1,400 °C for 3 h. Process II': 1400 °C for 8 h. Process III: annealing with tempering. After process II, the temperature was decreased to 1,250 °C and maintained for 1 h. Subsequently, the temperature was increased to 1,350 °C and maintained for 2 h. Process III': annealing at 1,400 °C for 3 h and 1,250 °C for 1 h. The figure is reprinted with permission from [71], Copyright © 2016 IOP Publishing

Table 3.1 Annealing time t_{ann} during the each step. The table is reprinted with permission from [71], Copyright © 2016 IOP Publishing

Process	1st annealing @1,000 (°C)	2nd annealing @1,400 (°C)	3rd annealing @1,250 (°C)	4th annealing @1,350 (°C)
I	1			
II	1	3		
II'	1	8		
III'	1	3	1	
III	1	3	1	2

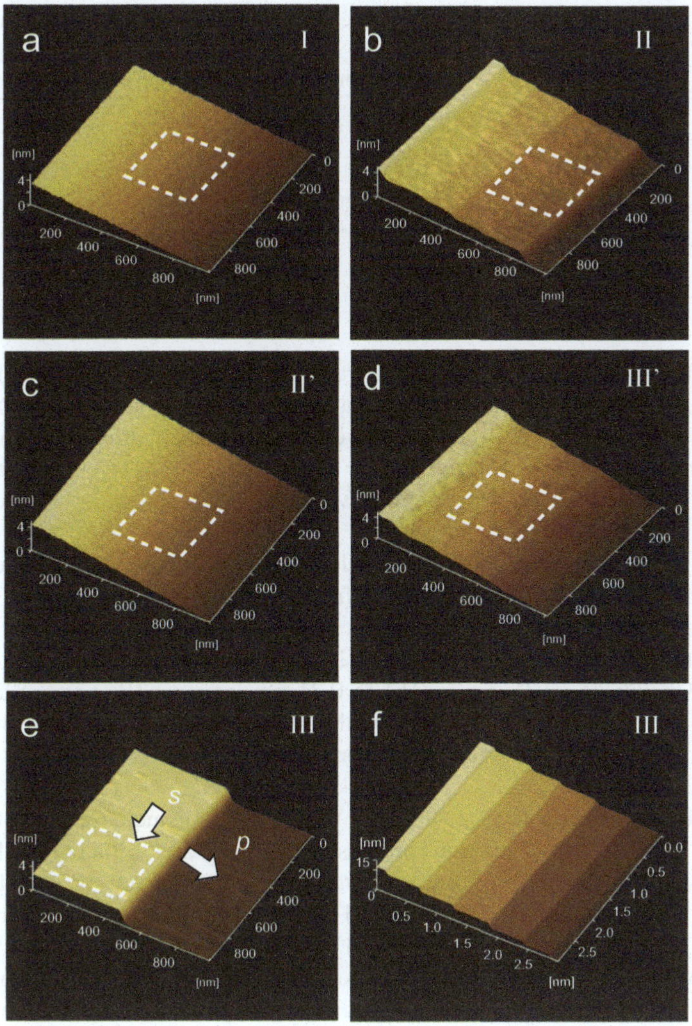

Fig. 3.52 AFM images of sapphire substrate. Annealing conditions: **a** process I, **b** process II, **c** process II', **d** process III', and **e** and **f** process III. *s* and *p* in **e** are the directions parallel and perpendicular to the atomic step. R_a was evaluated in the area indicated by the white dashed squares. The figures are reprinted with permission from [71], Copyright © 2016 IOP Publishing

(tempering sample). It was apparent of differences among each condition. To evaluate the difference, we obtained the surface roughness R_a. To eliminate measurement errors, we obtained five AFM images at different positions for each condition, one at center of the substrate, two at ±1.25 mm from the center in x-direction, and two at ±0.75 mm from the center in y-direction. Because there were large atomic step height in the surface obtained by processes II and III, we obtained R_a in 400×400 nm, so that the area did not include the large step edge. Table 3.2 shows the summary

Table 3.2 Averaged values of R_a. The table is reprinted with permission from [71], Copyright © 2016 IOP Publishing

Process	Average R_a (nm)	σ (nm)
I	0.379	0.023
II	0.152	0.057
II'	0.357	0.061
III'	0.268	0.118
III	0.051	0.02

of R_a of all process. From this table, it was found that we could obtain smaller R_a by higher T_{ann}: R_a decreased from 0.379 nm ($T_{\mathrm{ann}} = 1,000\,°\mathrm{C}$ (process I)) to 0.152 nm ($T_{\mathrm{ann}} = 1,400\,°\mathrm{C}$ (process II)); however, R_a increased by longer annealing time t_{ann}: R_a increased from 0.152 nm ($t_{\mathrm{ann}} = 3\,\mathrm{h}$ (process II)) to 0.357 nm ($t_{\mathrm{ann}} = 8\,\mathrm{h}$ (process II')). Smallest R_a (=0.051 nm) was obtained by annealing with tempering (process III).

In addition to the change of R_a, we evaluated the other parameters of the surface. We obtained the average step height, h_s from the AFM images (see Fig. 3.53a, b). From these results, we found that h_s increased from 0.27 nm (process I, $T_{\mathrm{ann}} = 1,000\,°\mathrm{C}$) to 1.63 nm (process II, $T_{\mathrm{ann}} = 1,400\,°\mathrm{C}$) as T_{ann} increased. h_s decreased from 1.63 nm (process II, $t_{\mathrm{ann}} = 3\,\mathrm{h}$) to 0.21 nm (process II', $t_{\mathrm{ann}} = 8\,\mathrm{h}$), as t_{ann} increased at $T_{\mathrm{ann}} = 1,400\,°\mathrm{C}$. However, although total t_{ann} of process III had much as that of process II', the value of h_s for process III (1.75 nm) was similar to that of process II (1.63 nm).

We also evaluated the terrace width, w. We found that w increased from 81.77 nm (process I, 1,000 °C) to 412.93 nm (process II, 1,400 °C) as T_{ann} increased. However, w decreased from 412.93 nm (process II, $t_{\mathrm{ann}} = 3\,\mathrm{h}$) to 56.66 nm (process II', $t_{\mathrm{ann}} = 8\,\mathrm{h}$), t_{ann} increased at $T_{\mathrm{ann}} = 1,400\,°\mathrm{C}$. However, the value of w for process III increased up to 508.93 nm, which derived from Fig. 3.52f.

As shown in Fig. 3.52f, we obtained periodic structure like a diffraction grating by the tempering process. However, the scanning stage of AFM system covers only several microns. Therefore, in order to evaluate the diffraction pattern from a larger area than AFM images, we obtained the diffraction pattern using an X-ray diffractometer system (SmartLab, Rigaku Corporation). The incident angle was 1° from the substrate, in the diffraction pattern measurement. Figure 3.54 shows the reflection patterns. There was no reflection angle dependences for pre-annealing (purple solid line and solid circles), process I (orange solid line and solid triangles), and process II', except for the mirror reflection same at 1° as the incident angle. However, there was first-order reflective diffracted light at $\theta = 1.78°$ (process II, solid blue curve in Fig. 3.54), $\theta = 1.69°$ (process III, solid red curve in Fig. 3.54), and 1.80° (process III', solid black curve in Fig. 3.54). These additional peaks with a broad width shown in Fig. 3.54 were originated from the reflection pattern of the periodical step structures.

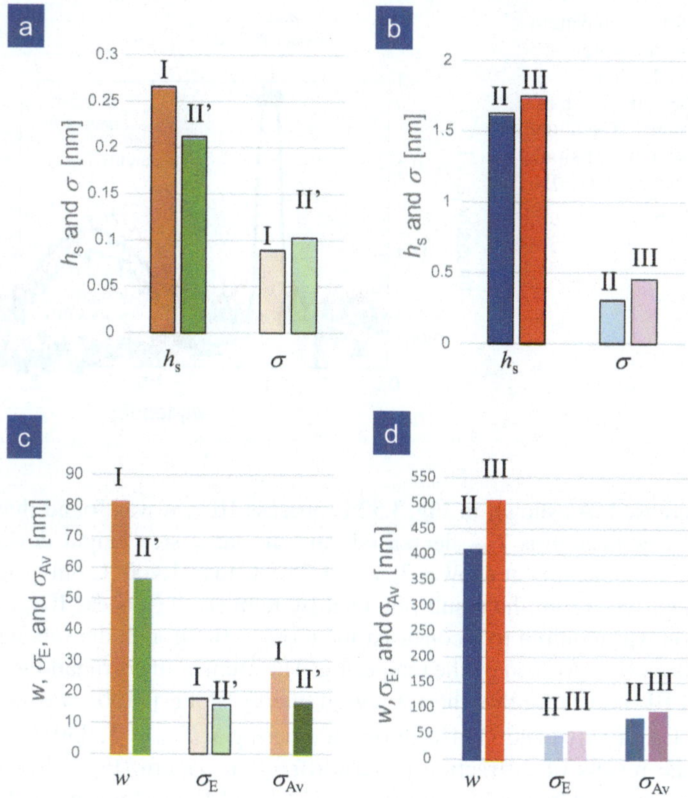

Fig. 3.53 Step height h_s and standard deviation σ for **a** processes I and II' and **b** processes II and III. Terrace width w, standard deviation of w in each terrace σ_E, and standard deviation of averaged w in each terrace σ_{AV} for **c** processes I and II' and **d** processes II and III. All data were averaged over five different positions; one at center of the substrate, two at ±1.25 mm from the center in x-direction, and two at ±0.75 mm from the center in y-direction. These figures are reprinted with permission from [71], Copyright © 2016 IOP Publishing

As reported in [58], the obtained value of $h_s = 0.27$ nm for process I (annealed at 1,100 °C) was good agreement with $h_u = 0.21$ nm [58]. In addition, the value of h_s increased as T_{ann} increased. Although we obtained atomic step structure, there were large corrugations along the step edge. The corrugations were originated from the sub-stable planes ($(2\bar{1}\bar{1}0)$ and $(\bar{1}2\bar{1}0)$) formation in addition to the main plane ($(1\bar{1}00)$) [61]. Both values of the σ_E (standard deviation of w in each terrace) and σ_{AV} (standard deviation of averaged w in each terrace) for process II' became smaller than that of process I (see Fig. 3.53c). Therefore, it was found that the uniform terrace with a smaller number of corrugations along $(1\bar{1}00)$ was obtained by the increase in T_{ann}.

Higher uniformity with fewer corrugations was obtained by the use of the tempering technique. As shown in Figs. 3.52b and 3.53a, the height deviation increased

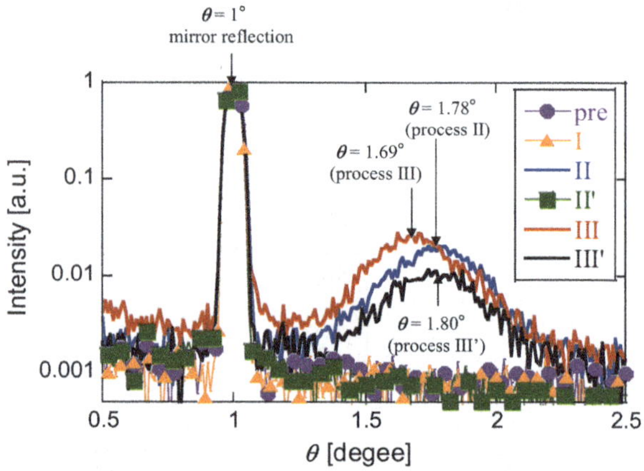

Fig. 3.54 Diffraction pattern from sapphire substrates. Normalized reflection spectra by the value taken at 1° incident angle. The figure is reprinted with permission from [71], Copyright © 2016 IOP Publishing

as T_{ann} increased. As shown in Fig. 3.52d (process III'), it was found that the step structures disappeared as T_{ann} decreased, and the large step height deviation was normalized by annealing at a lower T_{ann} of 1,250 °C than 1,400 °C only such as processes II and III'. From the result obtained by tempering (process III), the terrace deformation was required to reconstruct the terrace. In through the sub-stable state, wider terraces were formed by the tempering. The number of corrugations decreased in process III, indicating that the formation energy along $(1\overline{1}00)$ was more stable than that along $(2\overline{1}\overline{1}0)$ and $(1\overline{2}10)$ during tempering at $T_{\text{ann}} = 1,350\,°\text{C}$.

From the results of diffraction patterndiffraction pattern (Fig. 3.54), the small value of h_s as small as 0.2–0.3 nm (processes I and II') did not generate spectral peak. However, we observed peak of first-order reflective diffracted light at around 1.7–1.8° in processes II, III, and III'. The respective angles of first-order reflective diffracted light θ_1 and the value of terrace spacings d have a relation as follows

$$d = \frac{n\lambda}{\cos 1 - \cos \theta_1}. \tag{3.5}$$

By using the values of $n = 1$ for first-order reflective diffracted light and $\lambda = 0.15418\,\text{nm}$ for the Cu-Kα line), we obtained $d = 470.74$ (process II), $d = 541.49$ (process III), and $d = 451.96\,\text{nm}$ (process III'). As shown in Fig. 3.53d, the obtained terrace spacing ($d = 541.49\,\text{nm}$) for process III was in good agreement with the average spacing ($w = 508.93 \pm 57.40$) from AFM image. These results confirmed that the tempering process realized a millimeter-order scale uniformity of the atomic steps.

From the results of the step parameters in AFM images (Fig. 3.53), we evaluated the tilt angle of the surface ϕ (=$\tan^{-1}(h_s/w)$) was 0.19 (process I), 0.23 (process II), 0.21 (process II'), and 0.20° (process III). The obtained valued of ϕ for process III was same as the tilt angle (0.20°). Therefore, it was expected that the terrace formation by the tempering technique was influenced by the original substrate structure over the whole substrate in a millimeter scale.

In through the above evaluations, we concluded that the tempering technique resulted in the formation of uniform atomic step structure with wider terraces and higher steps with respect to the simple thermal annealing. Many group have reported uniform step structures with high steps and wide terrace widths by increasing T_{ann} and t_{ann}. However, due to the creation of an isolated step and step-by-step interactions at higher T_{ann} and longer t_{ann}, the step structure disappears [61, 75]. Therefore, the simple thermal annealing could not produce a uniform step structure with a large step height in a millimeter scale. In the tempering, the substrate was cooled down to 1,250 °C after annealing at 1,400 °C. The temperature of 1,250 °C is the threshold for step-step interactions [60]; therefore, it resulted in the deformation of the step structure formed at 1,400 °C. From the results shown in this section, such a deformation of the terrace resulted in the reconstruction of the terrace, and consequent wider terraces by increasing the temperature again, i.e., tempering. During the tempering, the substrate surface might be rather sub-stable stage, then it might be trigger to form the uniform terrace. The increase in the uniformity of a step structure over the large scale by the process III is good agreement with the tempering in chocolate production with pure-phase cacao [73].

References

1. T. Yatsui, K. Hirata, W. Nomura, Y. Tabata, M. Ohtsu, Realization of an ultra-flat silica surface with angstrom-scale average roughness using nonadiabatic optical near-field etching. Appl. Phys. B **93**(1), 55–57 (2008)
2. F. Brandenburg, R. Nagumo, K. Saichi, K. Tahara, T. Iwasaki, M. Hatano, F. Jelezko, R. Igarashi, T. Yatsui, Improving the electron spin properties of nitrogen-vacancy centres in nanodiamonds by near-field etching. Sci. Rep. **8**, 15847 (2018)
3. J.B. Burkholder, E.J. Bair, Potential energy parameters and shapes of the vibrational components of the 345-nm system of chlorine. J. Phys. Chem. **87**(11), 1859–1863 (1983)
4. R. Kullmer, D. Bäuerle, Laser-induced chemical etching of silicon in chlorine atmosphere. Appl. Phys. A **43**, 227–232 (1987)
5. T. Izawa, N. Inagaki, Materials and processes for fiber preform fabrication-vapor-phase axial deposition. Proc. IEEE **68**(10), 1184–1187 (1980)
6. R. Teki, A.J. Kadaksham, F. Goodwin, T. Yatsui, M. Ohtsu, Dressed-photon nanopolishing for extreme ultraviolet mask substrate defect mitigation. Proc. SPIE **8679**, 84–91 (2013)
7. Y. Hirabayashi, Development status of EUVL mask blank and substrate. Proc. SPIE **8166**, 989–994 (2011)
8. T. Yatsui, W. Nomura, F. Stehlin, O. Soppera, M. Naruse, M. Ohtsu, Challenges in realizing ultraflat materials surfaces. Beilstein J. Nanotechnol. **4**, 875–885 (2013)
9. D. Keilin, E. Hartree, Further observations on absorption spectra at low temperatures. Nature **165**, 504–505 (1950)
10. Z. Farooq, D.A. Chestakov, B. Yan, G.C. Groenenboom, W.J. van der Zande, D.H. Parker, Photodissociation of singlet oxygen in the UV region. Phys. Chem. Chem. Phys. **16**, 3305–3316 (2014)
11. T. Yatsui, W. Nomura, M. Ohtsu, Realization of ultraflat plastic film using dressed-photon-phonon-assisted selective etching of nanoscale structures. Adv. Opt. Technol. **2015**, 701802 (2015)

12. D.W. Allan, Statistics of atomic frequency standards. Proc. EEE **54**(2), 221–230 (1966)
13. T. Yatsui, W. Nomura, M. Naruse, M. Ohtsu, Realization of an atomically flat surface of diamond using dressed photon–phonon etching. J. Phys. D: Appl. Phys. **45**(47), 475302 (2012)
14. T. Yatsui, H. Saito, K. Nobusada, Angstrom-scale flatness using selective nanoscale etching. Beilstein J. Nanotechnol. **8**, 2181–2185 (2017)
15. P. Atkins, J. De Paula, J. Keeler, *Atkins' Physical Chemistry* (Oxford University Press, Oxford, 2017)
16. H. Chinen, K. Mawatari, Y. Pihosh, K. Morikawa, Y. Kazoe, T. Tsukahara, T. Kitamori, Enhancement of proton mobility in extended-nanospace channels. Angew. Chem. Int. Ed. **51**(15), 3573–3577 (2012)
17. S. Mononobe, M. Naya, T. Saiki, M. Ohtsu, Reproducible fabrication of a fiber probe with a nanometric protrusion for near-field optics. Appl. Opt. **36**(7), 1496–1500 (1997)
18. L. Wan, Y. Xu, Iodine-sensitized oxidation of ferrous ions under UV and visible light: the influencing factors and reaction mechanism. Photochem. Photobiol. Sci. **12**, 2084–2088 (2013)
19. G. Ondrey, N. van Veen, R. Bersohn, The state distribution of OH radicals photodissociated from H_2O_2 at 193 and 248 nm. J. Chem. Phys. **78**(6), 3732–3737 (1983)
20. M.P. Docker, A. Hodgson, J.P. Simons, Photodissociation of H_2O_2 at 248 nm: translational anisotropy and OH product state distributions. Chem. Phys. Lett. **128**(3), 264–269 (1986)
21. T. Yatsui, K. Hirata, Y. Tabata, Y. Miyake, Y. Akita, M. Yoshimoto, W. Nomura, T. Kawazoe, M. Naruse, M. Ohtsu, Self-organized near-field etching of the sidewalls of glass corrugations. Appl. Phys. B **103**(3), 527–530 (2011)
22. A. Gruber, A. Dräbenstedt, C. Tietz, L. Fleury, J. Wrachtrup, C. von Borczyskowski, Scanning confocal optical microscopy and magnetic resonance on single defect centers. Science **276**(5321), 2012–2014 (1997)
23. T. Staudacher, F. Shi, S. Pezzagna, J. Meijer, J. Du, C.A. Meriles, F. Reinhard, J. Wrachtrup, Nuclear magnetic resonance spectroscopy on a (5-nanometer)^3 sample volume. Science **339**(6119), 561–563 (2013)
24. F. Dolde, H. Fedder, M.W. Döherty, T. Nobauer, F. Rempp, G. Balasubramanian, T. Wolf, F. Reinhard, L.C.L. Hollenberg, F. Jelezko, J. Wrachtrup, Electric-field sensing using single diamond spins. Nat. Phys. **7**(6), 459–463 (2011)
25. G. Kucsko, P.C. Maurer, N.Y. Yao, M. Kubo, H.J. Noh, P.K. Lo, H. Park, M.D. Lukin, Nanometre-scale thermometry in a living cell. Nature **500**, 54–58 (2013)
26. L. Marseglia, J.P. Hadden, A.C. Stanley-Clarke, J.P. Harrison, B. Patton, Y.-L.D. Ho, B. Naydenov, F. Jelezko, J. Meijer, P.R. Dolan, J.M. Smith, J.G. Rarity, J.L. O'Brien, Nanofabricated solid immersion lenses registered to single emitters in diamond. Appl. Phys. Lett. **98**(13), 133107 (2011)
27. T. Yatsui, D. Takeuchi, S. Koizumi, K. Sato, K. Tsuzuki, T. Iwasaki, M. Hatano, T. Makino, M. Ogura, H. Kato, H. Okushi, S. Yamasaki, Polarization-controlled dressed-photon-phonon etching of patterned diamond structures. Phys. Status Solidi (a) **211**(10), 2339–2342 (2014)
28. H. Kato, T. Makino, M. Ogura, N. Tokuda, H. Okushi, S. Yamasaki, Selective growth of buried n^+ diamond on (001) phosphorus-doped n-type diamond film. Appl. Phys. Express **2**, 055502 (2009)
29. T. Yatsui, M. Kourogi, M. Ohtsu, Highly efficient excitation of optical near-field on an apertured fiber probe with an asymmetric structure. Appl. Phys. Lett. **71**(13), 1756–1758 (1997)
30. W. Nomura, M. Ohtsu, T. Yatsui, Nanodot coupler with a surface plasmon polariton condenser for optical far/near-field conversion. Appl. Phys. Lett. **86**(18), 181108 (2005)
31. T. Yatsui, W. Nomura, M. Ohtsu, Self-assembly of size- and position-controlled ultralong nanodot chains using near-field optical desorption. Nano Lett. **5**(12), 2548–2551 (2005)
32. T. Yatsui, M. Kourogi, M. Ohtsu, Plasmon waveguide for optical far/near-field conversion. Appl. Phys. Lett. **79**(27), 4583–4585 (2001)
33. T. Iwasaki, Y. Hoshino, K. Tsuzuki, H. Kato, T. Makino, M. Ogura, D. Takeuchi, T. Matsumoto, H. Okushi, S. Yamasaki, M. Hatano, Diamond junction field-effect transistors with selectively grown n^+-side gates. Appl. Phys. Express **5**(9), 091301 (2012)

34. T. Yatsui, H. Saito, K. Nishioka, B. Leuschel, O. Soppera, K. Nobusada, Effects of a power and photon energy of incident light on near-field etching properties. Appl. Phys. A **123**, 751 (2017)
35. K. Hasegawa, K. Yamada, R. Sasase, R. Miyazaki, A. Kikuchi, M. Yagi, Direct measurements of absolute concentration and lifetime of singlet oxygen in the gas phase by electron paramagnetic resonance. Chem. Phys. Lett. **457**(4), 312–314 (2008)
36. F. Stehlin, F. Wieder, A. Spangenberg, J.-M. Le Meins, O. Soppera, Room-temperature preparation of metal-oxide nanostructures by DUV lithography from metal-oxo clusters. J. Mater. Chem. C **2**, 277–285 (2014)
37. T. Wolf, P. Neumann, K. Nakamura, H. Sumiya, T. Ohshima, J. Isoya, J. Wrachtrup, Subpicotesla diamond magnetometry. Phys. Rev. X **5**, 041001 (2015)
38. C. Bradac, T. Gaebel, N. Naidoo, M.J. Sellars, J. Twamley, L.J. Brown, A.S. Barnard, T. Plakhotnik, A.V. Zvyagin, J.R. Rabeau, Observation and control of blinking nitrogen-vacancy centres in discrete nanodiamonds. Nat. Nanotechnol. **5**, 345–349 (2010)
39. L.P. McGuinness, Y. Yan, A. Stacey, D.A. Simpson, L.T. Hall, D. Maclaurin, S. Prawer, P. Mulvaney, J. Wrachtrup, F. Caruso, R.E. Scholten, L.C.L. Hollenberg, Quantum measurement and orientation tracking of fluorescent nanodiamonds inside living cells. Nat. Nanotechnol. **6**, 358–363 (2011)
40. R. Igarashi, Y. Yoshinari, H. Yokota, T. Sugi, F. Sugihara, K. Ikeda, H. Sumiya, S. Tsuji, I. Mori, H. Tochio, Y. Harada, M. Shirakawa, Real-time background-free selective imaging of fluorescent nanodiamonds in vivo. Nano Lett. **12**(11), 5726–5732 (2012)
41. L.-J. Su, C.-Y. Fang, Y.-T. Chang, K.-M. Chen, Y.-C. Yu, J.-H. Hsu, H.-C. Chang, Creation of high density ensembles of nitrogen-vacancy centers in nitrogen-rich type Ib nanodiamonds. Nanotechnology **24**(31), 315702 (2013)
42. R. Nagumo, F. Brandenburg, A. Ermakova, F. Jelezko, T. Yatsui, Spectral control of nanodiamond using dressed photon-phonon etching. Appl. Phys. A **121**(4), 1335–1339 (2015)
43. J.E. Tanner, E.O. Stejskal, Restricted self-diffusion of protons in colloidal systems by the pulsed-gradient, spin-echo method. J. Chem. Phys. **49**(4), 1768–1777 (1968)
44. R.C. Bansal, F.J. Vastola, P.L. Walker, Kinetics of chemisorption of oxygen on diamond. Carbon **10**(4), 443–448 (1972)
45. T. Yatsui, T. Tsuboi, M. Yamaguchi, K. Nobusada, S. Tojo, F. Stehlin, O. Soppera, D. Bloch, Optically controlled magnetic-field etching on the nano-scale. Light: Sci. Appl. **5**, e16054 (2016)
46. H. Ridaoui, F. Wieder, A. Ponche, O. Soppera, Direct ArF laser photopatterning of metal oxide nanostructures prepared by the sol-gel route. Nanotechnology **21**, 065303 (2011)
47. W. Nomura, T. Yatsui, Y. Yanase, K. Suzuki, M. Fujita, A. Kamata, M. Naruse, M. Ohtsu, Repairing nanoscale scratched grooves on polycrystalline ceramics using optical near-field assisted sputtering. Appl. Phys. B **99**, 75–78 (2010)
48. A. Krell, P. Blank, H. Ma, T. Hutzler, M.P.B. van Bruggen, R. Apetz, Transparent sintered corundum with high hardness and strength. J. Am. Ceram. Soc. **86**(1), 12–18 (2003)
49. A. Krell, P. Blank, H. Ma, T. Hutzler, M. Nebelung, Processing of high-density submicrometer Al_2O_3 for new applications. J. Am. Ceram. Soc. **86**(4), 546–553 (2003)
50. A. Ikesue, Polycrystalline Nd:YAG ceramics lasers. Opt. Mater. **19**(1), 183–187 (2002), *Photonic Materials for the 21st Century. Proceedings of the 2nd International Symposium on Laser, Scintillator and Nonlinear Optical Materials*
51. D. Graham-Rowee, Lasers for engine ignition. Nat. Photonics **2**, 515–516 (2008)
52. Z.-J. Liu, Y.G. Shen, Oscillating growth of surface roughness in multilayer films. Appl. Phys. Lett. **84**(25), 5121–5123 (2004)
53. E.J. Menke, Q. Li, R.M. Penner, Bismuth telluride (Bi_2Te_3) nanowires synthesized by cyclic electrodeposition/stripping coupled with step edge decoration. Nano Lett. **4**(10), 2009–2014 (2004)
54. F. Benabid, M. Notcutt, V. Loriette, L. Ju, D. Blair, X-ray induced absorption of high-purity sapphire and investigation of the origin of the residual absorption at 1064 nm. J. Phys. D **33**(10), 589 (2000)

55. K.F. MacDonald, V.A. Fedotov, S. Pochon, K.J. Ross, G.C. Stevens, N.I. Zheludev, W.S. Brocklesby, V.I. Emel'yanov, Optical control of gallium nanoparticle growth. Appl. Phys. Lett. **80**(9), 1643–1645 (2002)
56. J. Lim, T. Yatsui, M. Ohtsu, Observation of size-dependent resonance of near-field coupling between a deposited Zn dot and the probe apex during near-field optical chemical vapor deposition. IEICE Trans. Electron. **E88-C**(9), 1832–1835 (2005)
57. T. Yatsui, S. Takubo, J. Lim, W. Nomura, M. Kourogi, M. Ohtsu, Regulating the size and position of deposited Zn nanoparticles by optical near-field desorption using size-dependent resonance. Appl. Phys. Lett. **83**(9), 1716–1718 (2003)
58. M. Yoshimoto, T. Maeda, T. Ohnishi, H. Koinuma, O. Ishiyama, M. Shinohara, M. Kubo, R. Miura, A. Miyamoto, Atomic-scale formation of ultrasmooth surfaces on sapphire substrates for high-quality thin-film fabrication. Appl. Phys. Lett. **67**(18), 2615–2617 (1995)
59. L.P. Van, O. Kurnosikov, J. Cousty, Evolution of steps on vicinal (0001) surfaces of α-alumina. Surf. Sci. **411**(3), 263–271 (1998)
60. P.R. Ribič, G. Bratina, Behavior of the (0001) surface of sapphire upon high-temperature annealing. Surf. Sci. **601**(1), 44–49 (2007)
61. Y. Shiratsuchi, M. Yamamoto, Y. Kamada, Surface structure of self-organized sapphire (0001) substrates with various inclined angles. Jpn. J. Appl. Phys. **41**, 5719–5725 (2002)
62. Z.K. Tang, G.K.L. Wong, P. Yu, M. Kawasaki, A. Ohtomo, H. Koinuma, Y. Segawa, Room-temperature ultraviolet laser emission from self-assembled ZnO microcrystallite thin films. Appl. Phys. Lett. **72**(25), 3270–3272 (1998)
63. Y. Wang, S. Wang, S. Zhou, J. Xu, J. Ye, S. Gu, R. Zhang, Q. Ren, Effects of sapphire substrate annealing on ZnO epitaxial films grown by MOCVD. Appl. Surf. Sci. **253**(4), 1745–1747 (2006)
64. A. Tsukazaki, A. Ohtomo, T. Onuma, M. Ohtani, T. Makino, M. Sumiya, K. Ohtani, S.F. Chichibu, S. Fuke, Y. Segawa, H. Ohno, H. Koinuma, M. Kawasaki, Repeated temperature modulation epitaxy for p-type doping and light-emitting diode based on ZnO. Nat. Mater. **4**, 42–46 (2005)
65. H. Amano, N. Sawaki, I. Akasaki, Y. Toyoda, Metalorganic vapor phase epitaxial growth of a high quality GaN film using an AlN buffer layer. Appl. Phys. Lett. **48**(5), 353–355 (1986)
66. S. Nakamura, The roles of structural imperfections in InGaN-based blue light-emitting diodes and laser diodes. Science **281**(5379), 956–961 (1998)
67. M. Sumiya, K. Yoshimura, T. Ito, K. Ohtsuka, S. Fuke, K. Mizuno, M. Yoshimoto, H. Koinuma, A. Ohtomo, M. Kawasaki, Growth mode and surface morphology of a GaN film deposited along the N-face polar direction on c-plane sapphire substrate. J. Appl. Phys. **88**(2), 1158–1165 (2000)
68. M. Yoshimoto, K. Yoshida, H. Maruta, Y. Hishitani, H. Koinuma, S. Nishio, M. Kakihana, T. Tachibana, Epitaxial diamond growth on sapphire in an oxidizing environment. Nature **399**, 340–342 (1999)
69. T. Tsukamoto, T. Ogino, Morphology of graphene on step-controlled sapphire surfaces. Appl. Phys. Express **2**, 075502 (2009)
70. H. Ago, K. Imamoto, N. Ishigami, R. Ohdo, K.-I. Ikeda, M. Tsuji, Competition and cooperation between lattice-oriented growth and step-templated growth of aligned carbon nanotubes on sapphire. Appl. Phys. Lett. **90**(12), 123112 (2007)
71. T. Yatsui, K. Kuribara, T. Sekitani, T. Someya, M. Yoshimoto, Temperature-modulated annealing of c-plane sapphire for long-range-ordered atomic steps. J. Phys. D **49**, 115302 (2016)
72. C.E. Clarkson, T. Malki, Alternation in long-chain compounds. Part II. An X-ray and thermal investigation of the triglycerides. J. Chem. Soc. 666–671 (1934)
73. C. Loisel, G. Keller, G. Lecq, C. Bourgaux, M. Ollivon, Phase transitions and polymorphism of cocoa butter. J. Am. Oil Chem. Soc. **75**, 425–439 (1998)
74. S.E. Demina, E.N. Bystrova, M.A. Lukanina, V.M. Mamedov, V.S. Yuferev, E.V. Eskov, M.V. Nikolenko, V.S. Postolov, V.V. Kalaev, Numerical analysis of sapphire crystal growth by the kyropoulos technique. Opt. Mater. **30**(1), 62–65 (2007)
75. E.D. Williams, R.J. Phaneuf, J. Wei, N.C. Bartelt, T.L. Einstein, Thermodynamics and statistical mechanics of the faceting of stepped Si(111). Surf. Sci. **294**(3), 219–242 (1993)

Chapter 4
Near-Field Chemical Reactions and Excitations

4.1 Introduction

It is required to use the power from the sun efficiently for a sustainable development. However, we need UV photon energy to excite materials, including water splitting [1], carbon dioxide reduction [2]. Researcher has been exploring new materials with narrower band gap [3, 4].

4.2 Near-Field Assisted Chemical Reactions

The electron transfer efficiency between the electrode and solution depends strongly on the rearrangement energy of the solvent (L) [5, 6]. To overcome the L (see Fig. 4.1), we use thermal heating (process (a) in Fig. 4.1). The use of a catalyst results in reduction of the L value. Although platinum (Pt) is the one of the best catalysts to reduce the L, because Pt is a rare material, it is required to reduce the Pt amount to be used. To reduce the usage amount, Pt nanoparticles can be utilized so that the surface area increased and it results in saving the Pt usages [7, 8], in which the electron transfer efficiency does not change from that of a flat surface. Therefore, it is required to improve the electron transfer efficiency itself of Pt [9].

We applied the ONF to realize higher efficiency of electron transfer [10]. The ONF can couple with the intermediate state of electron ((b_1) process in Fig. 4.1b). Thus, the ONF could realize multi-step excitation via the intermediate state, and the resultant energy exceeds the L. Consequently, the solvent has energy exceeding the rearrangement energy. Therefore, the ONF excitation (Fig. 4.1b) has higher efficiency of the electron transfer than that of the conventional photo-excitation (Fig. 4.1a)

To validate the electron transfer of the ONF excitation, we evaluated current generation in ferricyanide solution (0.01 M $K_3[Fe(CN)_6]$ with 1 M KNO_3 aqueous solution.) [11] using a three-electrode setup (see Fig. 4.2) $Fe(CN)_6^{3-}$ is stable ion in the solution. It exchanges electrons ($Fe(CN)_6^{3-} + e^- \rightleftharpoons Fe(CN)_6^{4-}$), in which no

T. Yatsui, *Nanophotonic Chemical Reactions*, Nano-Optics and Nanophotonics,
https://doi.org/10.1007/978-3-030-42843-3_4

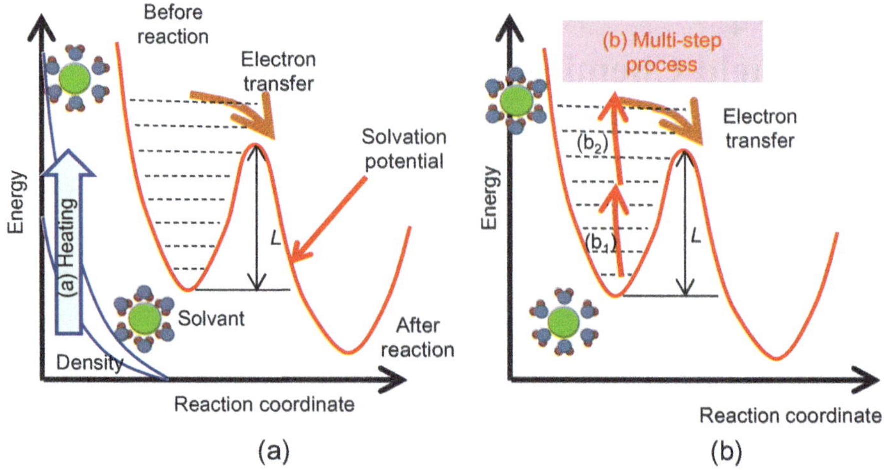

Fig. 4.1 Schematic of the electron transfer. **a** Single-step excitation by conventional thermal process. **b** Multi-step excitation by the near-field process. These figures are reprinted from [10] under the terms of the Creative Commons Attribution license

Fig. 4.2 Schematic of the experiment setup. These figures are reprinted from [10] under the terms of the Creative Commons Attribution license

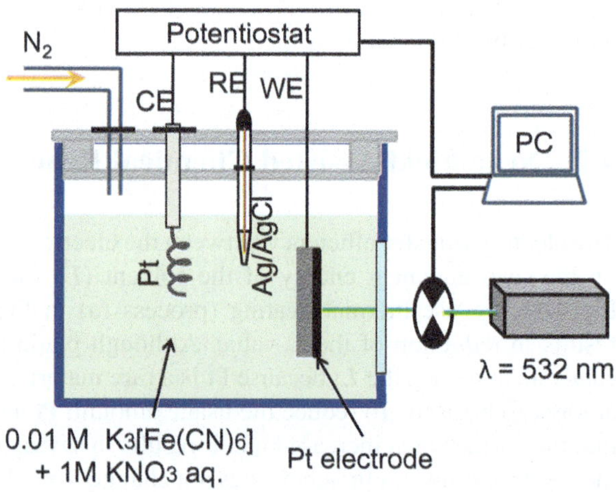

chemical change occurs. Therefore, the generated current can be originated from the number of electrons exchange. During the experiment, we supplied nitrogen gas so that it reduces the dissolved oxygen in the solution (Fig. 4.3).

Because the absorption edge wavelength of $Fe(CN)_6^{3-}$ is around 450 nm (Fig. 4.2) [12], we used the 532-nm light. Thus, it can be excluded that the generated current in this experiment, the observed current generated using 532-nm light, is originated from the ONF excitation.

We used a platinum black electrode (Fig. 4.4b) [13, 14] as a working electrode (WE). By comparing the AFM image of platinum film (Fig. 4.4a), the platinum black

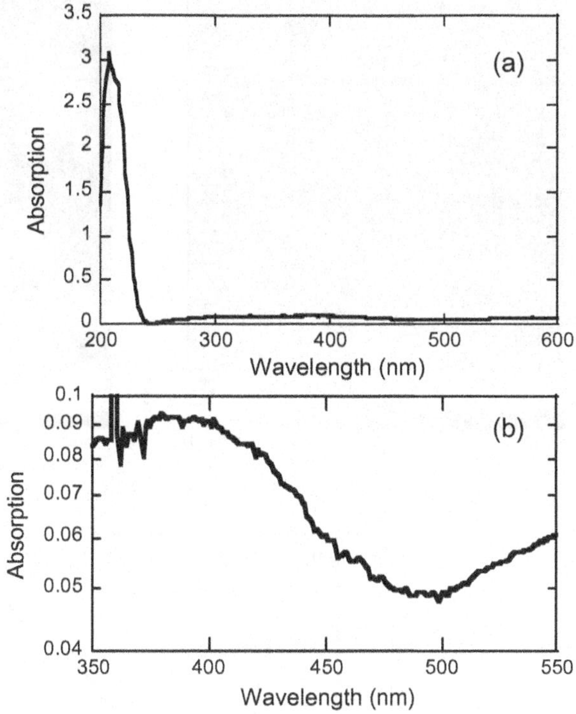

Fig. 4.3 Absorption spectra of $Fe(CN)_6^{3-}$ solution (**b**) are magnified spectra of **a**. These figures are reprinted from [10] under the terms of the Creative Commons Attribution license

electrode has structures in nanometer scale (Fig. 4.4a). From the AFM image of the platinum black electrode (Fig. 4.4b), the particle size was found to be in the range from 30 to 70 nm. The thickness was 4.2 μm.

From the dark cyclic voltammogram of ferricyanide solution (Fig. 4.5a), the current density was constant in the range between −0.5 and +0.2 V for the potential voltage (vs. the reference electrode (RE)). Based on the result, we performed the experiment at −0.3 V (vs. RE), so that we could detect the effect of photo-excitation only. We irradiated laser for 100 s with 100 s separation (see Fig. 4.5b). We compared the average current density between on and off periods. Figure 4.5c shows the laser power dependence of the average current density, indicating that the current density increased for on as laser power increased; however, it unchanged for off.

The solid circles in Fig. 4.6a show the normalized current density ($\Delta i / i$), in which Δi is the difference between on and off averaged current density. We estimated the heat generation in a Pt black electrode. Because we deposited Pt black on a sapphire substrate, we evaluated the heat generation in the model Pt (thickness of L_1)/sapphire (thickness of L_2) (see Fig. 4.6b). In the calculation model, we consider the following assumptions:

- The maximum temperature increase by taking into account that light power was absorbed into heat at the Pt film interface.

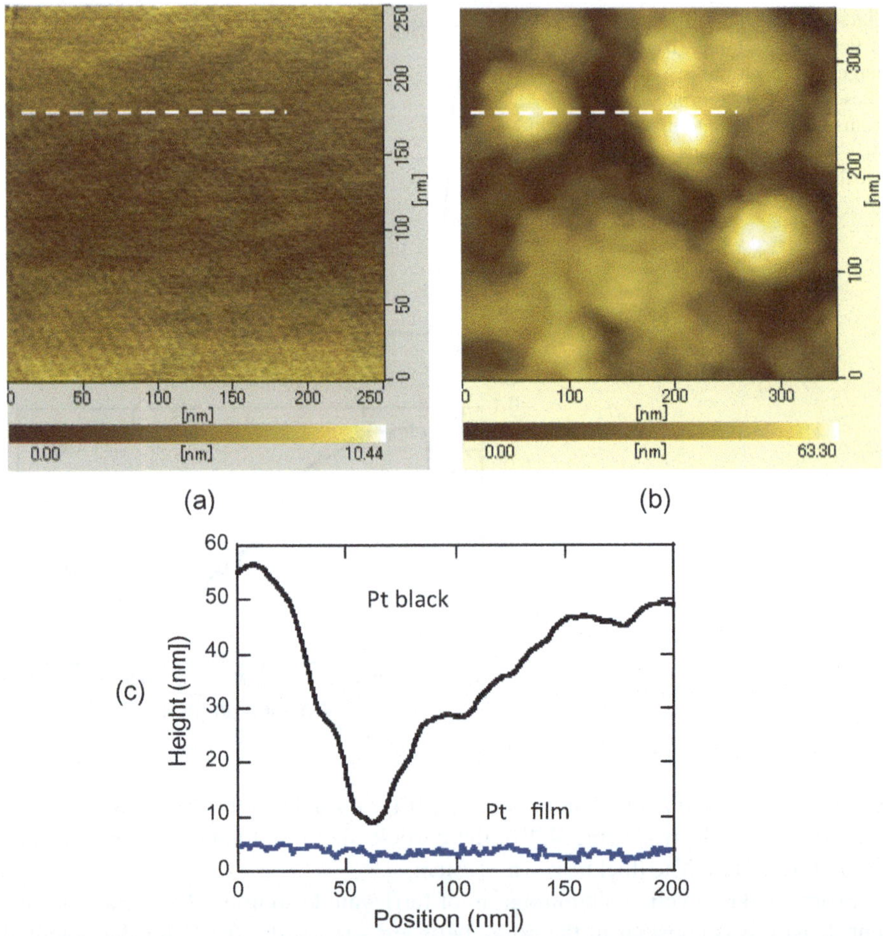

Fig. 4.4 AFM images of **a** platinum film and **b** platinum black. **c** Cross-sectional profiled of white dashed line in **a** and **b**. These figures are reprinted from [10] under the terms of the Creative Commons Attribution license

- No heat transfer into the solution.
- The temperature at the interface between the sapphire (back side) and the solution reached room temperature.
- Calculation model was one-dimensional conductive heat transfer model (see Fig. 4.6b). In the model, the heat transferred from the Pt film to the sapphire substrate.

The increase of temperature ΔT between the front and back sides is given by the following equation [5]

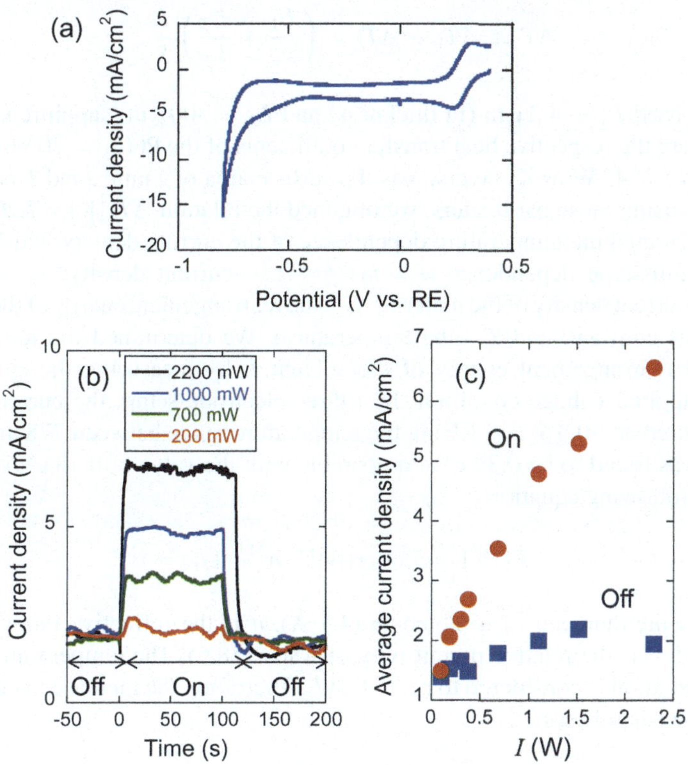

Fig. 4.5 **a** Cyclic voltammogram of ferricyanide solution without irradiation. **b** The temporal evaluation of the current density at −0.3 V (vs. RE). **c** The laser power dependence of the average current density. On: closed circles, off: closed squares. These figures are reprinted from [10] under the terms of the Creative Commons Attribution license

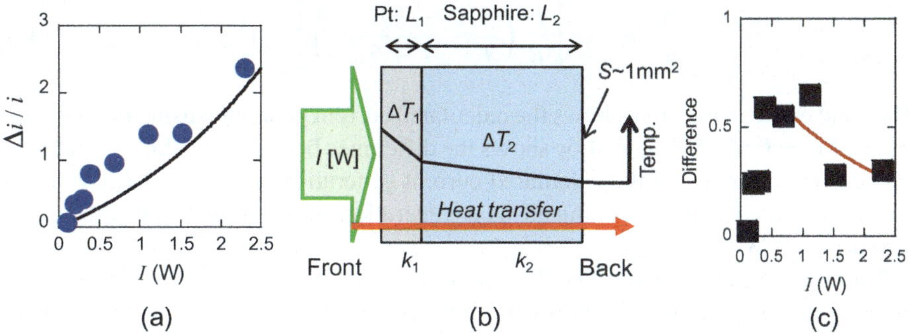

Fig. 4.6 **a** Incident light power dependence of the normalized current density. **b** Schematic of one-dimensional model of the electrode structure consisting of Pt (thickness L_1) and sapphire substrate (thickness L_2). **c** The incident light power dependence of the difference between the normalized current density (solid circles in **a**) and calculated current generation using the L value of 0.48 eV (solid curve in **a**). The red solid line: the fitted curve for $I > 0.5$ W. These figures are reprinted from [10] under the terms of the Creative Commons Attribution license

$$\Delta T \equiv \Delta T_1 + \Delta T_2 = \left(\frac{L_1}{k_1} + \frac{L_2}{k_2}\right)\frac{I}{S} \tag{4.1}$$

where we used $L_1 = 4.2\,\mu\text{m}$ (Pt thickness) and $L_2 = 300\,\mu\text{m}$ (sapphire substrate), k_1 and k_2 are the respective heat transfer coefficients of the Pt ($k_1 = 70\,\text{W/m K}$) and sapphire ($k_2 = 42\,\text{W/m K}$) layers, S is the surface area of $1\,\text{mm}^2$, and I is the laser power. By using those parameters, we obtained the relation ΔT [K] = 7.20 I [W].

We calculated the temperature dependence of the current density which follows the Arrhenius-type dependence ($i = i_0 e^{-L/RT}$; i—current density; i_0—the quasi exchange current density of the material; L—the rearrangement energy of the solvent; R—the gas constant; and T—the temperature). We determined the temperature-dependent rearrangement energy of the solvent L by measuring the current at a constant applied voltage condition. In a three-electrode setup, the current density was measured at −0.3 V (vs. RE) in the temperature range between 278 and 328 K. Thus, L was found to be 0.39 eV, comparable with the value of 0.48 eV calculated using the following equation

$$L[\text{eV}] = 3.60/\text{r}[\text{Å}](1/\text{n}^2 - 1/_{\text{r}}) \tag{4.2}$$

where r is the diameter of a molecule (4.1 Å), n is the refractive index of water (1.33), and ϵ_r is the relative permittivity of water (78.5). The temperature increase by the irradiation is considered to be $T + \Delta T$. Therefore, the increase in current can be described as follows;

$$i + \Delta i = i_0 e^{-L/R(T+\Delta T)} \tag{4.3}$$

With the relation ΔT [K] = 7.20 I [W], the maximum current increase with laser heating is obtained as:

$$\frac{\Delta i}{i} = \exp\left[\frac{L}{R}\left(\frac{1}{T} - \frac{1}{T + 7.2I}\right)\right] - 1 \tag{4.4}$$

The solid curve in Fig. 4.6a shows the calculated current generated using the obtained value of $L = 0.48\,\text{eV}$. Figure 4.6c shows the difference between the obtained normalized current density and the calculated current generation with $L = 0.48\,\text{eV}$. These results indicated that laser-induced heat generation was dominant at higher laser power than 2 W.

Based on the comparison, we decreased the irradiation power to avoid the laser-induced heating. Figure 4.7 shows the light power dependence of the normalized current density at low laser power. Dashed lines in Fig. 4.7 show the calculated value of the heat generated in a Pt black electrode using (4.4), indicating that the heat generation was negligible with respect to that of the ONF process. It was found that higher-order dependencies on I appeared with increasing current density. The blue solid curves in Fig. 4.7 were fit by the second-order function $\Delta i/i = aI + bI^2$. From the comparison between the experimental and calculated values, it is reasonable to

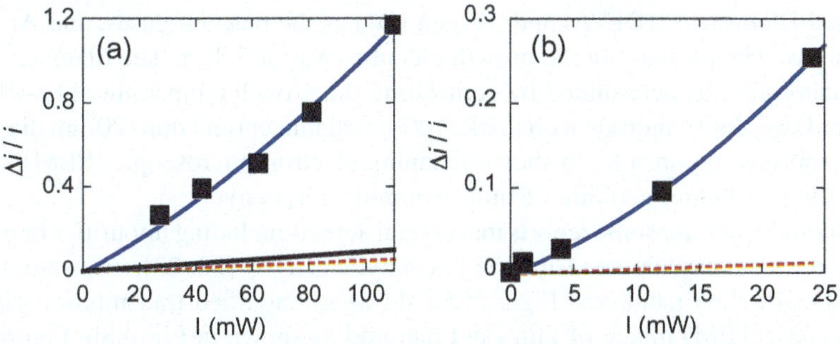

Fig. 4.7 Irradiation power dependence of current density. **a** $I < 110$ mW, **b** $I < 25$ mW. Solid squares: the observed current generation for platinum black. Solid black line: the observed current generation for platinum film. The blue solid curves: the calculated current increase by **a** $\Delta i/i = 7.3 \times 10^{-3} I + 3.0 \times 10^{-5} I^2$ and **b** $\Delta i/i = 5.8 \times 10^{-3} I + 2.0 \times 10^{-5} I^2$. The red dashed curves: the calculated heat generation using (4.4). These figures are reprinted from [10] under the terms of the Creative Commons Attribution license

the observed data excited by the light irradiation follow the two-step activation of electron transfer. Those results support the postulate that the electron transfer was originated from the two-step excitation owing to the ONF generated at the platinum black electrode with nanoscale structures. It was also supported that the fact of greater increase of the current increase in comparison with the platinum film (red dashed curves in Fig. 4.7a).

4.3 Near-Field Assisted Water Splitting

To realize water splitting using the sunlight, TiO_2 is one of the best materials because of high efficiency. However, since TiO_2 is a wide-band gap material (band gap energy of 3.2 eV), it requires higher photon energy to generate the carriers than the band gap energy. Recent development of the material on the band engineering has made it possible to realize water splitting using narrow-band gap materials [15, 16]. To realize higher efficiency of water splitting, we utilized the ONF [17].

4.3.1 ZnO Nanorod with Nanoscale Diameter

We used ZnO nanorods as the ONF source as well as photocatalytic electrode, because it has nanometer-scale apex. Using metal-organic vapor phase epitaxy (MOVPE), we synthesized the ZnO nanorods on a sapphire (0001) substrate [18]. To use the ZnO nanorod as an electrode, the sapphire substrate was coated with an indium tin oxide (ITO) thin film (200 nm) before the ZnO growth. In the MOVPE process,

We used Diethylzinc (DEZn) and oxygen (O_2) as the reactant gases, and Ar as the carrier gas. The pressure in the growth chamber was at 5 Torr. The diameter of the ZnO nanorods was determined by controlling the growth temperature [19–21]. We prepared two ZnO nanorods with thick (100-nm diameter) and thin (20-nm diameter) ZnO nanorods. Figure 4.8a, b shows scanning electron microscopic (SEM) images of the thick (100 nm) and thin (20 nm) nanorods, respectively.

Although there are some reports that crystal defects including impurities or oxygen deficiencies increase in the visible photocatalytic activity [22, 23], we examined the properties of ZnO nanorods. Figure 4.8c shows a magnified transmission electron microscopy (TEM) image of thin ZnO nanorod as shown in Fig. 4.8b. Figure 4.8d, e shows the cross-sectional profiles along the white dashed lines of d-d′ and e-e′ in Fig. 4.8c, respectively. Those results confirmed that the fabricated ZnO nanorods had single crystallinity with lattice spacings of 0.503 and 0.298 nm. These values are consistent with crystalline wurtzite ZnO with lattice constants of 0.521 nm (*c* axis) and 0.325 nm (*a* axis) [24]. Thus, it was confirmed that nanorods grew along the *c*-axis. As a reference ZnO material, We used a commercially available ZnO bulk substrate by hydrothermal synthesis and a sputtered ZnO film as references. From the AFM images of the reference samples, the surface roughness R_a of the ZnO bulk substrate and the sputtered ZnO film were found to be 0.21 nm and 6.63 nm, respectively.

Figure 4.9c shows the cathodoluminescence (CL) spectra of the different types of ZnO electrodes. These spectra have similar spectral shape, having a single peak energy at 3.3 eV. The emission peak was originated from the band edge emission of ZnO (indicated as $I_B E$ in Fig. 4.9c). If ZnO has the crystal defect levels due to oxygen vacancies, it results in the green emission with a photon energy of $h\nu \sim 2.6$ eV [25–27]. To evaluate the quality of the ZnO samples, we obtained the ratios of the CL intensity at $h\nu = 2.6$ eV to that at $h\nu$ 3.3 eV. It was found to be 0.02 (the bulk substrate, the sputtered film, and thick (100 nm) nanorods), and less than 0.001 (the thin (20 nm) nanorods). Therefore, we could exclude the possibility of multi-step excitation due to impurities.

Figure 4.10 shows the experimental setup of the water splitting. We used Pt wire as the counter electrode and Ag/AgCl as the reference electrode. The laser spot was sin 1 mm. We used 0.1 M NaOH solution as the electrolyte. To decrease in the measurement error, we performed the repeatedly irradiation for 20 s with 20 s intervals and obtained the averaged values. We used a potentiostat for the electrochemical measurements. Hydrogen and oxygen were generated at the Pt counter electrode and oxygen at the ZnO working electrode [28].

We evaluated dependences of the electrode potential for 4.66 eV exceeding 3.3 eV of the ZnO band gap energy (Fig. 4.11a). We set an input power at 300 μW. We could observe the flat band for all electrodes at below −0.7 V (vs. Ag/AgCl) [29, 30]. From the results, we set 0.5 V (vs. Ag/AgCl) of the electrode potential. Figure 4.11b shows the photo-current dependence as a function of the 4.66-eV-laser power with an electrode potential of 0.5 V (vs. Ag/AgCl). From these results, we obtained linear dependence on the incident laser power. We fitted the best-fit lines in Fig. 4.11b. Although each sample had different surface areas, the observed values

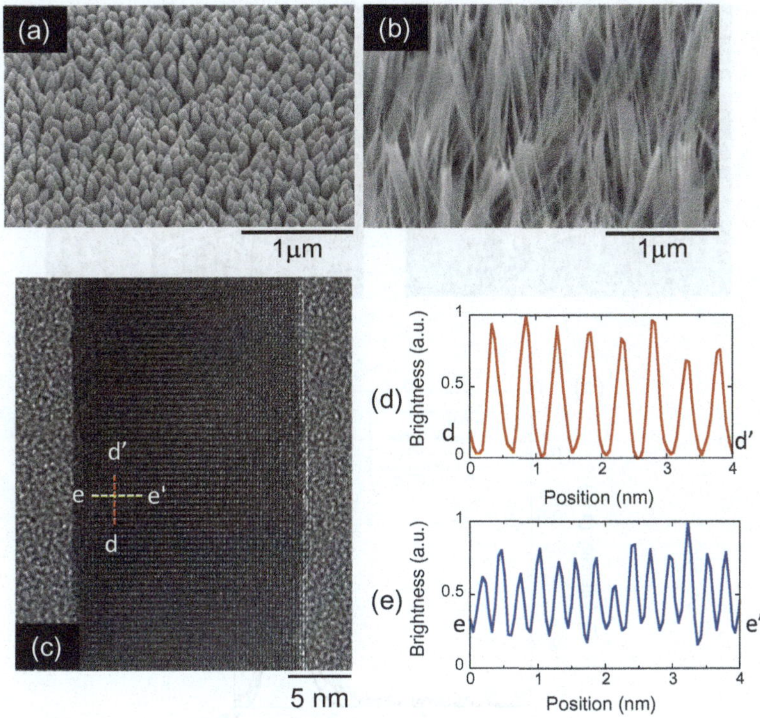

Fig. 4.8 SEM images of **a** thick (100 nm) nanorods and **b** thin (20 nm) nanorods. **c** TEM image of the thin (20 nm) nanorods. Cross-sectional profiles along the dashed lines **d** d-d′ and **e** e-e′ in **c**, respectively. These figures are reprinted from [17] under the terms of the Creative Commons Attribution license

of photo-currents for all samples had similar amount. The results indicated that we could eliminate the effect of surface areas for photo-current generation. In addition, the incident photon exceeding the band gap energy, we could exclude the ONF effect.

We evaluated the photo-current dependence on the laser power with $h\nu = 2.62$, 2.33, and 1.85 eV (see Fig. 4.12a–c), below the band gap energy (3.3 eV). Although we detected small photo-current from the bulk ZnO substrate (closed squares in Fig. 4.12), we obtained greater photo-currents for both thick (100 nm, closed triangles) and thin (20 nm, closed circles) nanorods electrodes. From results shown in Fig. 4.12a–c, we obtained a linear photo-current dependence on laser power. We obtained 86 times increase in the photo-current of thick nanorods in comparison with that of bulk substrate in 2.62-eV-excitation (Fig. 4.12a).

We evaluated the incident photon-to-current conversion efficiency (IPCE) as a function of the incident photon energy (see Fig. 4.13). The IPCE value of the thick (100 nm) and thin (20 nm) nanorods has much larger than that of bulk electrode at all other excitation levels. In addition, we found that the IPCE value decreased as the excitation photon energy decrease.

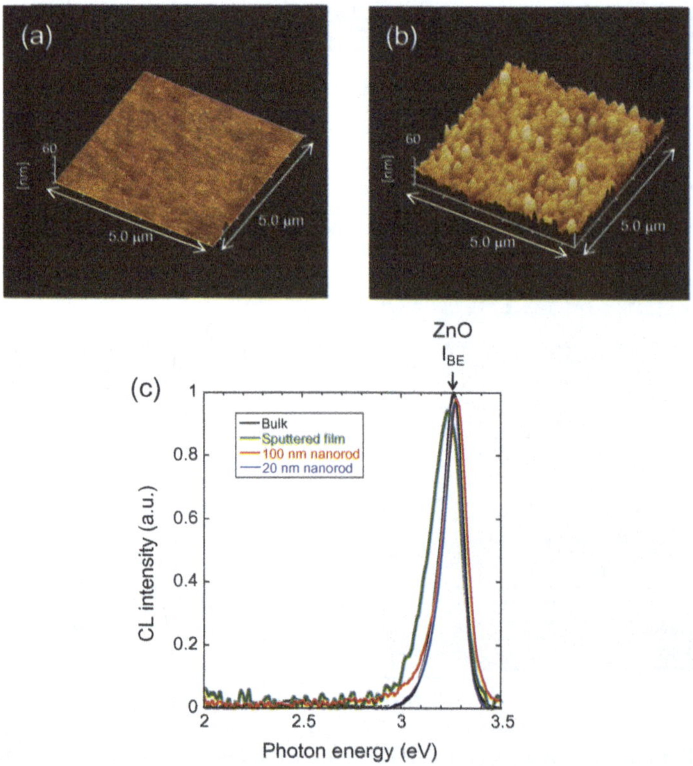

Fig. 4.9 AFM images of the **a** ZnO bulk substrate and **b** sputtered ZnO film. **c** Solid curves correspond to the CL spectrum of the bulk substrate (black), the spectrum of the 100 nm nanorods (red), and spectrum of the 20 nm nanorods (blue). The figure is reprinted from [17] under the terms of the Creative Commons Attribution license

Fig. 4.10 Schematic of the photocatalytic water splitting. ZnO: the working electrode, Ag/AgCl: the reference electrode, and Pt: the counter electrode. The figure is reprinted from [17] under the terms of the Creative Commons Attribution license

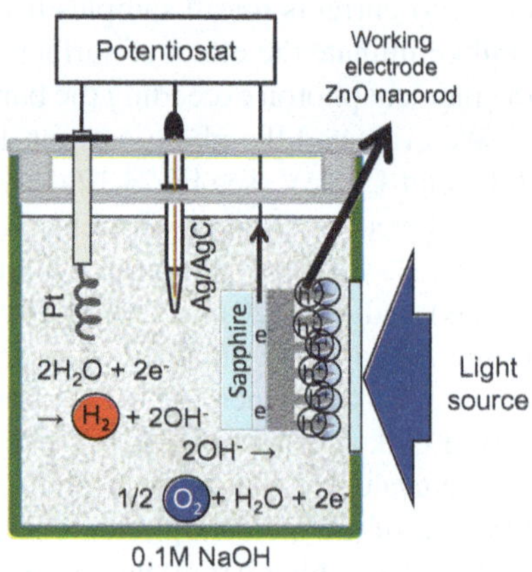

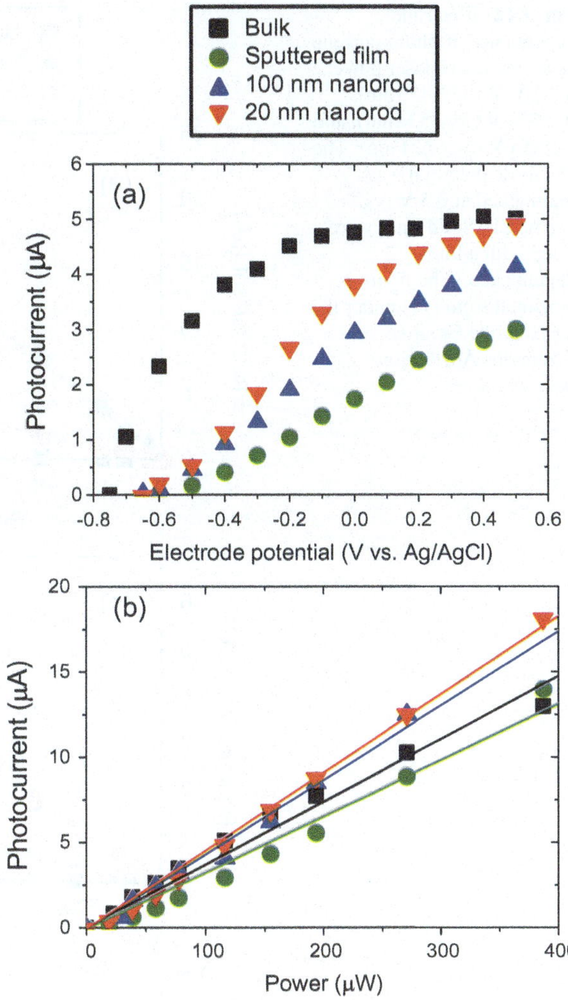

Fig. 4.11 Photo-current dependence on **a** electrode potential and **b** 4.66-eV-light power. Electrode potential was 0.5 V (vs. Ag/AgCl) in **b**. Solid lines in **b** were fitted lines with a linear dependence. The figure is reprinted from [17] under the terms of the Creative Commons Attribution license

4.3.2 TiO_2 Nanorod with Nanoscale Structures

Because the non-uniformity of the generated optical field strongly depends on the nanometric structures, we need to control the shape of nanostructures to enhance the ONF effect [31]. There have been developing various techniques for 1D nanostructures fabrication, among which glancing angle deposition (GLAD) is one of the candidates to meet the requirement to fabricate aligned nanorod with well-controlled rod diameter, density, and shape by changing deposition parameters [32, 33]. Herein, we demonstrate the use of radio frequency magnetron sputtering deposition with GLAD technique to fabricate nanorod TiO_2 with nanostructures.

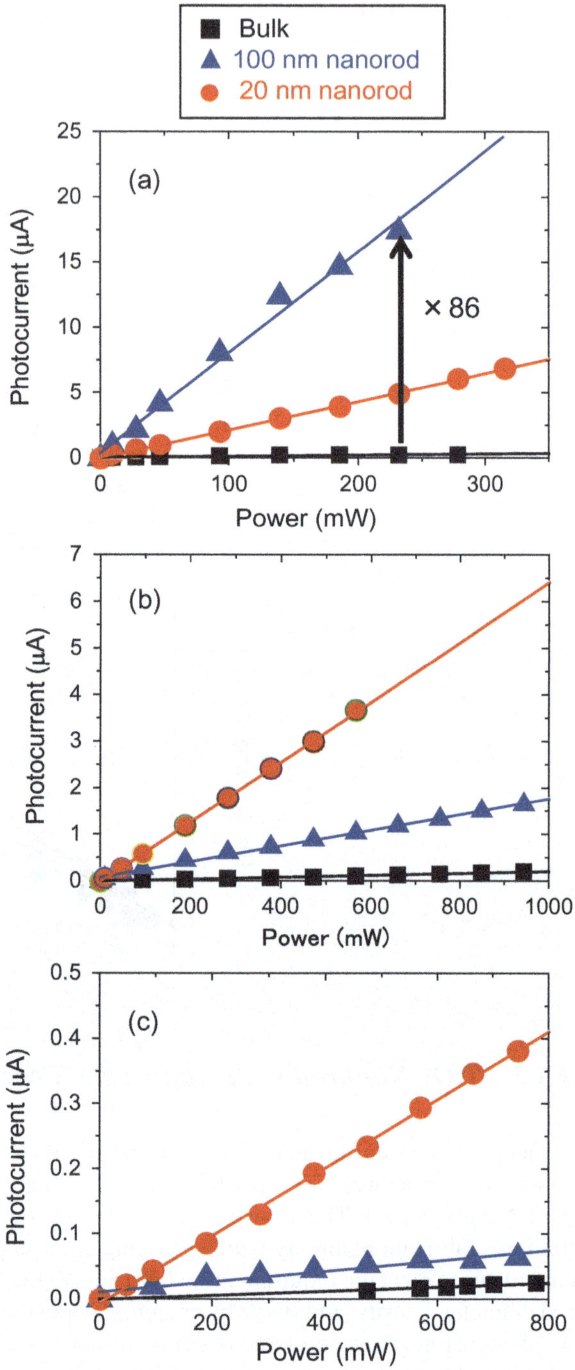

Fig. 4.12 Electrode dependence of photo-current as a function of laser power. **a** 2.62 eV ($\lambda = 473\,\text{nm}$), **b** 2.33 eV ($\lambda = 532\,\text{nm}$), and **c** 1.85 eV ($\lambda = 671\,\text{nm}$). The electrode potential was maintained at 0.5 V (vs. Ag/AgCl). Solid lines were fitted with a linear dependence. The figure is reprinted from [17] under the terms of the Creative Commons Attribution license

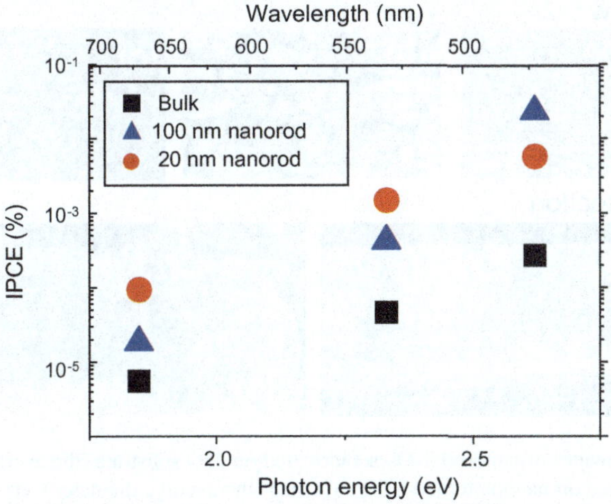

Fig. 4.13 The power dependency of photo-response current measured by PEC method at 0.5 V applied bias (vs. SCE) in 0.5M $NaClO_4$ of **a** under 365 nm irradiation and **b** 488 nm irradiation of 130 nm nanorod (blue circle), 65 nm nanorod (green square), and P-25 flat film (red triangle). The photo-response of fabricated nanorod TiO_2 samples is comparative to the P-25 nanoparticles film; however, **b** under 488 nm irradiation, the photo-current in nanorods samples is more than 2 orders of magnitude larger photo-currents, compared to that of P-25 film. A liner dependency is observed in a wide range of light power from $10^2 \sim 10^6$ mWcm^{-2}. The figure is reprinted from [17] under the terms of the Creative Commons Attribution license

Figure 4.14 shows the SEM images TiO_2 nanorods. To grow TiO_2 nanorods with less than 100 nm in diameter, we coated flat ITO (indium tin oxide) on glass substrate. While, to grow TiO_2 nanorods with larger than 100 nm, we used pyramid-like nanopatterns on the FTO (F:doped Tin Oxide) substrate which acts as nucleation sites for the incoming target flux. For the reference purpose, we fabricated the flat film made of Degussa P-25 nanoparticles by spin-coating.

We found that we obtained rough surfaces with nanostructures on the larger nanorod (130 nm), while the smooth surfaces on the smaller (65 nm) nanorod sample.

We evaluated the photocatalytic performance of TiO_2 photo-anode by photo-electrochemical (PEC) measurement, in which we used a three-electrode setup in 0.5M $NaClO_4$ electrolyte and 0.5 V (vs. Ag/AgCl) applied bias. Figure 4.15a, b shows the power dependency of photo-current under 365-nm light (UV) and 488-nm light (visible) irradiation, respectively. Under UV irradiation, the nanorod TiO_2 films had a similar performance as P-25 (see Fig. 4.15a). In contrast, under visible (488 nm) irradiation, 130-nm nanorod sample had more than 2 orders of magnitude enhancement of photo-currents (see Fig. 4.15b). We obtained the large enhancement only in the visible region by introducing nanorod structures that excludes the possibility of the photo-current enhancement due to enhanced surface area and improved charge separation and transport properties. We found the linear dependency in a wide range of $10^{-2} \sim 10^{-5}$ mWcm^{-2} and the slope of equal to 1.0, which excludes the possi-

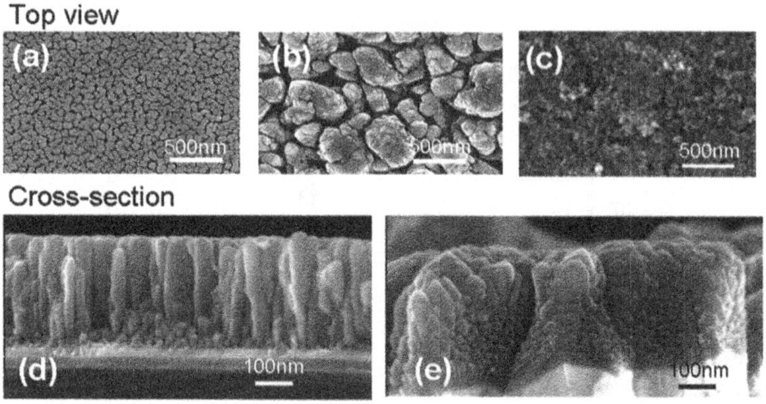

Fig. 4.14 SEM images of nanorod TiO_2 **a** fabricated on ITO substrate (the average diameter was 65 nm), **b** fabricated on nanopatterned FTO substrate (the average diameter were 130 nm). **c** SEM image of top view of flat film made of Degussa P-25 nanoparticles. Cross-sectional SEM images of **d** 65-nm nanorod sample and **e** 130-nm nanorod sample. These figures are reprinted with permission from [31], Copyright © 2011 American Institute of Physics

bility of the multi-step excitation. It is also worthy to note that the photo-current has been acquired in the range of 10^{-5} An even under visible irradiation eliminates the factors of thermal effect or background noise during the measurement. If we attribute the enhancement of visible response to the ONF-assisted model, the observed linear relation can be explained as the creating of electron-hole pairs was governed only by the dipole-forbidden transition of the first step.

4.4 Near-Field Assisted CO_2 Reduction

4.4.1 Direct CO_2 Reduction

Carbon dioxide (CO_2) reduction is one of the most important challenges in realization of a sustainable development. Because the dissociation energy of the CO_2 molecule is as much as 7.47 eV ($\lambda_{dis} = 166$ nm) [34], the increase of CO_2 molecule would be crucial problem for global warming. Thus, the sun is not good light source to dissociate the CO_2 molecule. To dissociate the CO_2 molecule, people try to dissociate by alternative method, including plasma [35], magnetite [36], and CH_4 [37]. However, the dissociation efficiency is not high and/or they require a vacuum chamber. Therefore, to realize higher efficiency of CO_2 reduction, we developed near-field assisted CO_2 reduction [38].

We used ZnO nanorods as the ONF source for near-field assisted CO_2 reduction. We prepared ZnO nanorods using two-temperature growth on a sapphire substrate

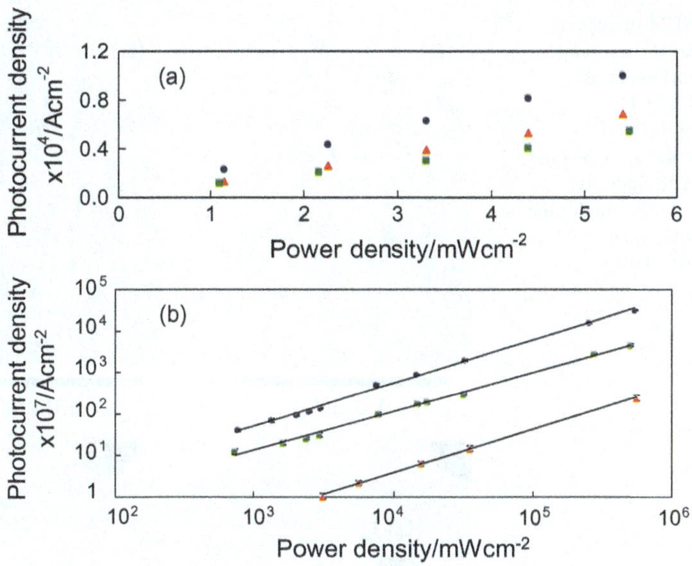

Fig. 4.15 The incident power dependency of photo-current density. The photo-current was measured by PEC method at 0.5 V applied bias (vs. SCE) in 0.5M NaClO4. **a** Under 365-nm light (UV) irradiation and. **b** Under 488-nm light (visible) irradiation. Blue circles: 130 nm nanorod, green squares: 65 nm nanorod, and red triangles: P-25 flat film. A liner dependency is observed in a wide range of light power from 10^2 to 10^{-6} mWcm^{-2}. These figures are reprinted with permission from [31], Copyright © 2011 American Institute of Physics

[19]. We used thick (100 nm) and thin (20 nm) ZnO nanorods as show in SEM images (see Fig. 4.16a, b).

To evaluate the amount of CO_2 molecules, we detected the absorption spectrum of CO_2 in a chamber. We used a halogen lamp for the detection of the absorption spectrum. First, we measured the intensity (I_0) of the vacuum, i.e., without CO_2 gas. Second, we measured that of CO_2 (I). Then, we obtained absorbance, defined by $A = \ln I_0/I$.

Figure 4.17a shows the absorption spectrum of CO_2.

As described in details in [5], CO_2 has four independent modes, including symmetric stretching (ν_1), antisymmetric stretching (ν_3), and two perpendicular bending modes (ν_2). A mode is expressed as $\nu_1\nu_2 l\nu_3 n$, in which l is the angular momentum quantum number and n is the nth component in states linked by rovibrational interaction.

The measured absorption spectra were good agreement with the high-resolution transmission molecular absorption (HITRAN) database [39]. Therefore, we considered that the spectrum was originated from the antisymmetric stretching mode. Here, we considered two modes at 1434.19 (mode A; 00001 → 00031) and 1441.94 nm (mode B (01101 → 01131), where the antisymmetric stretching (ν_3) number changes. Figure 4.17b shows the band diagram of the A and B modes.

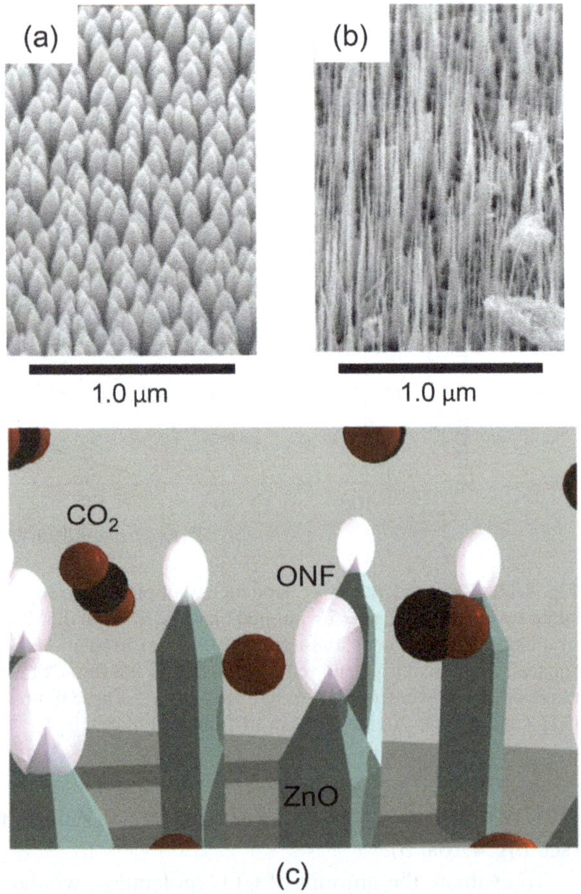

Fig. 4.16 SEM images of ZnO **a** thick (100 nm) and **b** thin (20 nm) nanorods. **c** Schematic of CO_2 reduction by the ONF generated at the apex of ZnO nanorods. The figure is reprinted from [38] under the terms of the Creative Commons Attribution license

First, we obtained time dependence of modes A and B. Figure 4.18a shows the time dependences by the irradiation with $\lambda = 213$ nm. As shown in this figure, we could not observe any change in the case without nanorods, as shown in dashed lines (see Fig. 4.18a). However, the peak intensity for both modes A and B decreased in similar manner. Second, we obtained time dependence of modes A and B by the irradiation with $\lambda = 266$ nm (Fig. 4.18b). Now, we could not observe any change in case without nanorods, as shown in dashed lines (see Fig. 4.18b. In this condition, the peak intensity for mode A decreased, on the contrary that for mode B increased with increased as the exposure time increased.

The absorption wavelength of the CO_2 molecule was from 120 to 200 nm. As reported in details in [40–42], the deep UV (DUV) absorption is originated from the electronic transitions from the ground state ($^1\Sigma_g^+$ to singlet ($^1B_2$) and triplet ($^3B_2$) states, as shown in Fig. 4.19. Slanger et al. reported that the 147-nm light irradiation results in the CO_2 dissociation [40, 41], and Wayne evaluated the quantum yield of

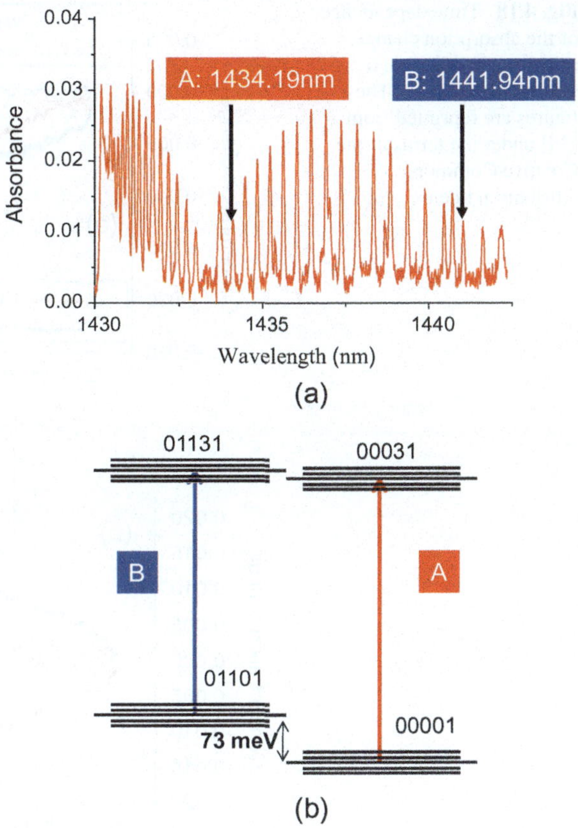

Fig. 4.17 **a** Typical absorption spectrum of CO_2. **b** Energy band diagram. Mode A at 1434.19 nm: asymmetric vibrational transition of 00001→00031. Mode B at 1441.94 nm: hot band 01101→01131. The figure is reprinted from [38] under the terms of the Creative Commons Attribution license

CO and O (^{1}D) production as 1.0 [42]. One of the first dissociation paths considered as:

$$CO_2 + h\nu \rightarrow CO(^1\Sigma_g{}^+) + O(^1D). \tag{4.5}$$

where the potential energy corresponds to λ <166 nm (see Fig. 4.19). The other possible dissociation path considered as:

$$CO_2 + h\nu \rightarrow CO(^1\Sigma_g{}^+) + O(^3P), \tag{4.6}$$

where a transition occurs from the singlet state ($^1B_2$) to the triplet state ($^3B_2$) with a potential energy corresponds to $\lambda = 184.9$ nm. There were some reports the experimental confirmation of the dissociation process [43, 44].

From those results, we found the followings:

- CO_2 molecules could not be dissociated without nanorods, i.e., no ONF (as shown in the dashed lines in Fig. 4.18a, b).

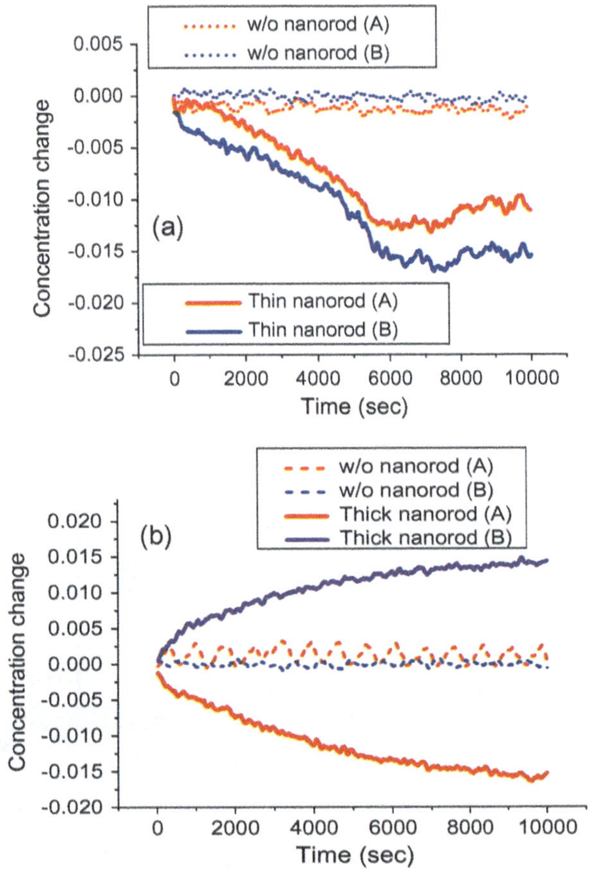

Fig. 4.18 Time dependence of the absorption change. **a** 213-nm irradiation. **b** 266-nm irradiation. These figures are reprinted from [38] under the terms of the Creative Commons Attribution license

- By the 213-nm light irradiation with ZnO nanorods, the absorption peaks decreased (as shown in the solid lines in Fig. 4.18a, b). This result confirmed that the ONF could dissociate CO_2 molecules.
- By the 266-nm light irradiation with ZnO nanorods, the peak intensity of A mode decreased, while that of B mode increased. This result is also explained by the near-field effect.

4.4.2 CO_2 Reduction with Metal Complex

To realize higher reduction efficiency, the use of metal complex has been investigated, such as Re [45], Ru [46], and Cu [47]. We investigated the ONF effect on the metal complex [48] using Re(bpy-COOH)$(CO)_3$Cl (bpy-COOH = 2,2′-bipyridine-4,4′-dicarboxylic acid), abbreviated as Re(bpy-COOH) [49].

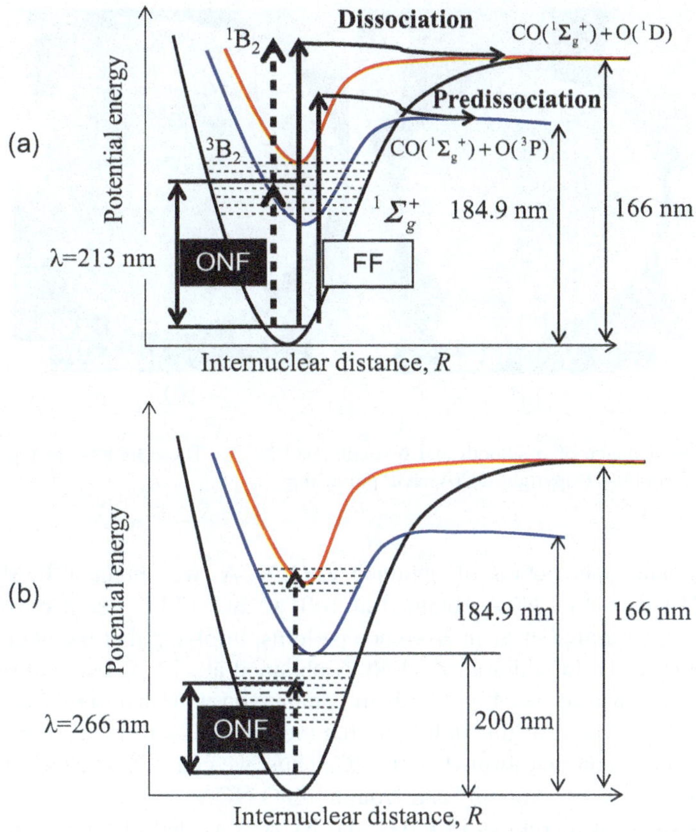

Fig. 4.19 Dissociation of the CO_2 molecule in an ONF-assisted process for **a** 213 nm. **b** 266 nm. FF: Far-field transition with a single photon. ONF: Optical near-field transition via vibrational levels. The figure is reprinted from [38] under the terms of the Creative Commons Attribution license

To generate the ONF, we used ZnO nanocrystalline aggregates (NCA). Based on the [50], we synthesized ZnO NCA by the hydrolysis of zinc salt in polyol medium.

Zinc acetate dihydrate (0.001 mol) was added to diethylene glycol (100 mL) while stirring (700 rpm). During the process, the solution was heated at 160 °C (smooth ZnO NCA) and 170 °C (rough ZnO NCA). The colloidal solution was concentrated by centrifugally separating the aggregates from the solvent. After removing the supernatant, the precipitation was re-dispersed in ethanol. Finally, we obtained the ZnO NCA. We obtained ZnO NCA with different surface roughness with different heating temperature at 160 °C and 170 °C. Figure 4.20a, b corresponds to the SEM images synthesized at 160 °C and 170 °C, respectively. From those SEM images, we found that the ZnO NCA synthesized at 170 °C had a much higher number of small nanocrystals.

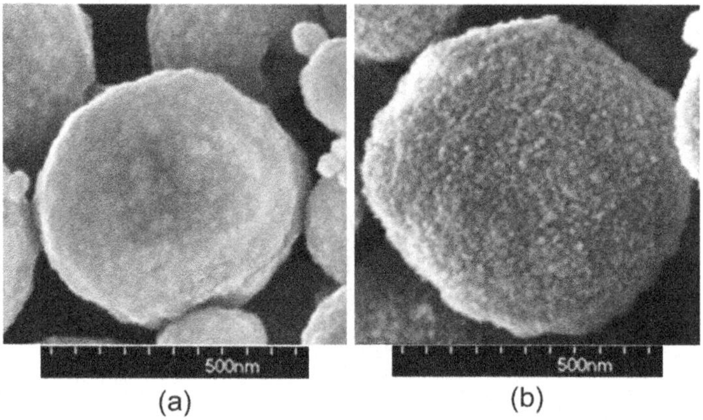

Fig. 4.20 SEM image of **a** smooth and **b** rough ZnO NCA. These figures are reprinted with permission from [48], Copyright © 2019 IOP publishing

To evaluate more details of obtained ZnO NCA, we obtained TEM images. Figure 4.21a–d is ZnO NCA obtained at 160 °C and 170 °C, respectively. From Fig. 4.21a, c, we found clear interference patterns, implying that the obtained ZnO has single crystals. In addition, ZnO NCA obtained at 170 °C had nanoscale protrusions at the surface (see Fig. 4.21d) in comparison with that obtained at 160 °C (see Fig. 4.21b). Those results indicated that the ZnO NCA formed at 170 °C (ZnO rougher surface than that formed at 160 °C. Thus, we called ZnO NCA obtained at 160 °C and 170 °C as "smooth" and "rough" ZnO NCA.

As described in literatures [50, 51], the NCA consisted of monodisperse ZnO nanocrystallines. The average diameter of NCA was around 12 nm. We evaluated the size distributions of rough ZnO NCA from SEM images (see Fig. 4.22d), indicating that the obtained size distributions were comparable to the reported value of 12 nm. Furthermore, it is reported that there was no apparent size difference depending on the heating temperatures as the heating temperature increases. In addition, the ZnO NCA could not keep the spherical shape when the heating temperature increases. From those results and reported information, the surface roughness of rough ZnO NCA was due to the detachment of the ZnO nanocrystalline with 12 nm in diameter. Therefore, rough ZnO NCA should have the surface roughness in the range of 12 nm.

We also evaluated the ZnO NCA by the optical property. Figure 4.23 shows the absorption spectra of the ZnO NCA. We used spectrometer (JASCO FP-8200ST). From the obtained spectra, we confirmed that the obtained ZnO NCA have an absorption edge at 380 nm, indicating the obtained ZnO NCA had high-quality crystallinity [52, 53].

As for the Re complex, we used Re(bpy-COOH)$(CO)_3$Cl (bpy-COOH = 2,2′-bipyridine-4,4′-dicarboxylic acid), abbreviated as Re(bpy-COOH) [49]. As described in [54], first, $Re(CO)_5Cl$ was dissolved in methanol. The same amount of 2,2′-Bipyridine-4,4′-dicarboxylic acid was added to the solution, and the reaction mixture

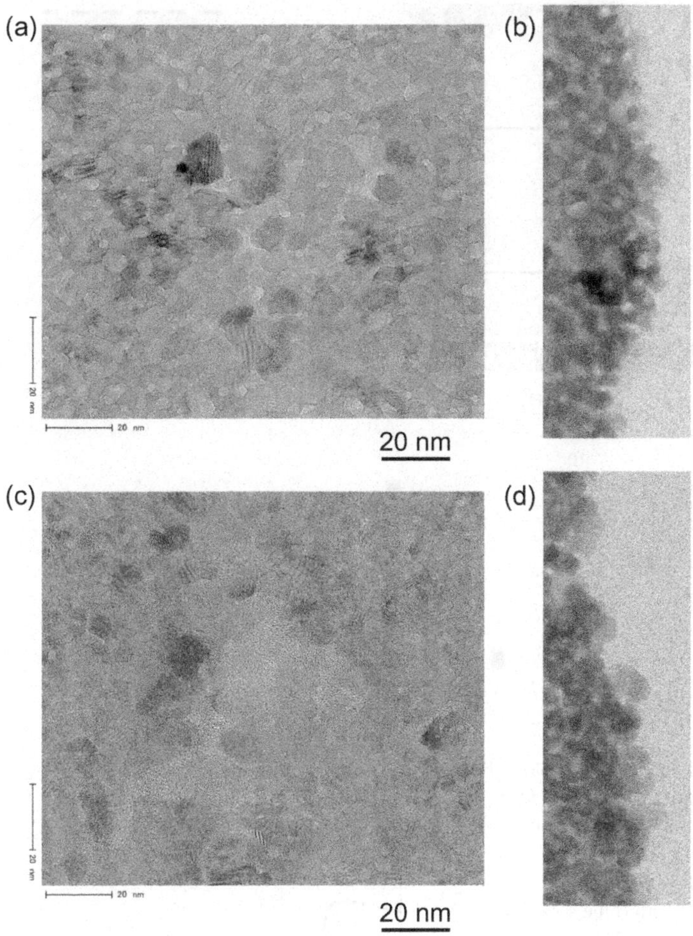

Fig. 4.21 TEM images. **a** and **b** Smooth ZnO NCA obtained at 160 °C. **c** and **d** rough ZnO NCA obtained at 170 °C. **a** and **c** were taken at the center of NCA. **b** and **d** were taken at the edge of NCA

was stirred under reflux over night. Then, we obtained Re(bpy-COOH). Figure 4.24 shows the instruction of the process. Then, to deposit Re(bpy-COOH) onto the ZnO NCA, Re(bpy-COOH) was dissolved in acetonitrile. Then, the Re(bpy-COOH) solution was mixed with the ZnO NCA solution, and the mixture was stand still over night. Then, Re(bpy-COOH) was deposited onto ZnO NCA, in which a proton was removed from the carboxyl group (deprotonation) (see Fig. 4.24).

To validate the deposition of Re(bpy-COOH), we obtained the Fourier transform infrared (FTIR) spectra. We used the KBr tablet method with Shimadzu IRPrestige-21. Figure 4.25 shows the FTIR spectra. We also obtained the spectra of Re(bpy-COONa) in which H was exchanged by Na as a reference. In Fig. 4.25,

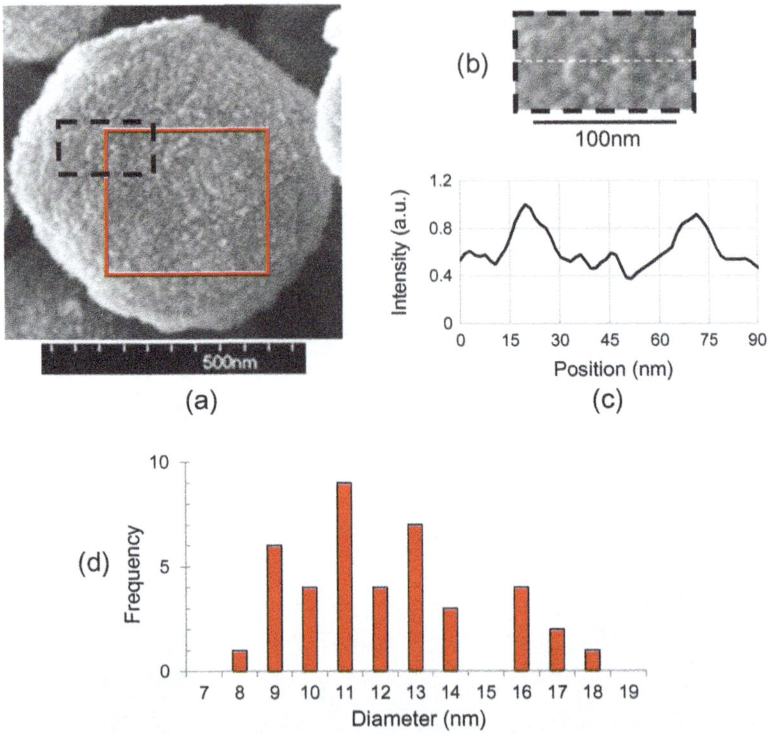

Fig. 4.22 **a** SEM image of rough ZnO NCA. **b** Magnified image of **a**. **c** Cross-sectional profile along the white dashed line in **b**. **d** The size distribution of ZnO nanocrystallines. These figures are reprinted with permission from [48], Copyright © 2019 IOP publishing

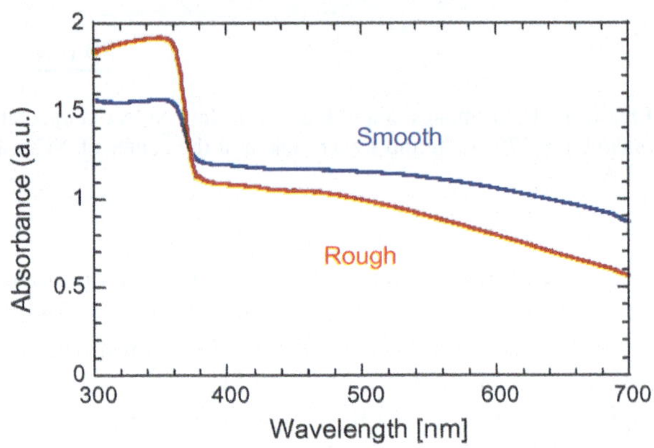

Fig. 4.23 The absorption spectra of smooth (blue solid line) and rough (red solid line) ZnO NCA. These figures are reprinted with permission from [48], Copyright © 2019 IOP publishing

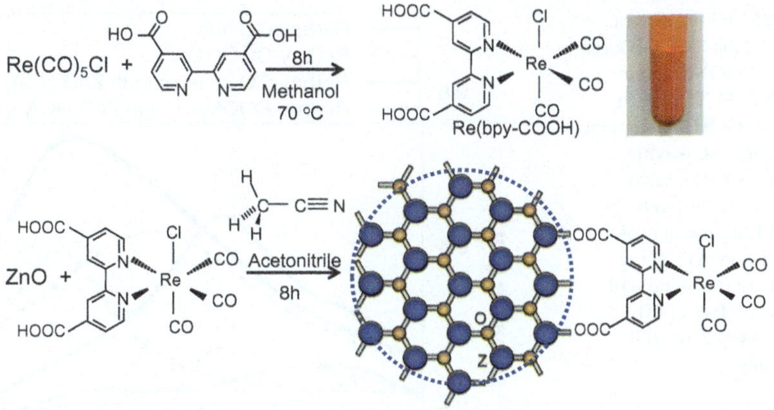

Fig. 4.24 Schematic of the Re complex deposits onto ZnO NCA. These figures are reprinted with permission from [48], Copyright © 2019 IOP publishing

the absorption spectrum of Re(bpy-COOH) is shown as the black solid curve. To see the changes in the FTIR spectra in details, we fitted the curves with Lorentzian peaks as shown in dashed curves in Fig. 4.25. The main peak at 2044.0 cm^{-1} corresponds to the stretching mode of the CO ligands [55]. The Re(bpy-COONa) spectrum has a main peak at 2031.1 cm^{-1} (see red solid curve in Fig. 4.25). The Re(bpy-COONa) spectrum shifted to a lower wave number by 12.9 cm^{-1} in comparison with Re(bpy-COOH). The shift is caused because the COO^- from the COONa group is a weaker electron-withdrawing group than that of the COOH group. Thus, Re for COO^- has larger electron density than Re for COOH, and the bond order of CO becomes small for COO^-, which results in a lower wave number of Re(bpy-COONa), in comparison with Re (bpy-COOH) [54]. The spectra of both smooth (blue solid curve in Fig. 4.25, main peak at 2026.4 cm^{-1}) and rough (green solid curve in Fig. 4.25, main peak at 2033.1 cm^{-1}) ZnO NCA with Re(bpy-COOH) have similar lower shifts of the main peak in comparison with that for Re(bpy-COOH). These results support that Re(bpy-COOH) was deposited to the ZnO NCA by the deprotonation of the carboxyl group.

We compared the absorption spectra to check the effect of ONF excitation on the Re complex deposited onto ZnO NCA. Because ZnO NCA had different size, we measured the diffuse reflection spectra to eliminate the light scattering effect, depending on the difference of the aggregate size [56]. We used a spectrophotometer (JASCO V-670) to obtain the diffuse reflection spectra. Figure 4.26a shows the Kubelka-Munk plots of the diffuse reflection spectra of Re(bpy-COOH) (solid black curve), Re(bpy-COONa) (solid green curve), and Re(bpy-COOH) deposited onto smooth ZnO NCA (solid blue curve) [56]. In these spectra, we observed two main peaks at 300 nm and at 400 nm. Those two peaks are originated from the $\pi - \pi^*$ transition and metal-to-ligand charge transfer (MLCT) transition, respectively [55]. To find the change of spectra in more details, we fitted them with three Lorentzian

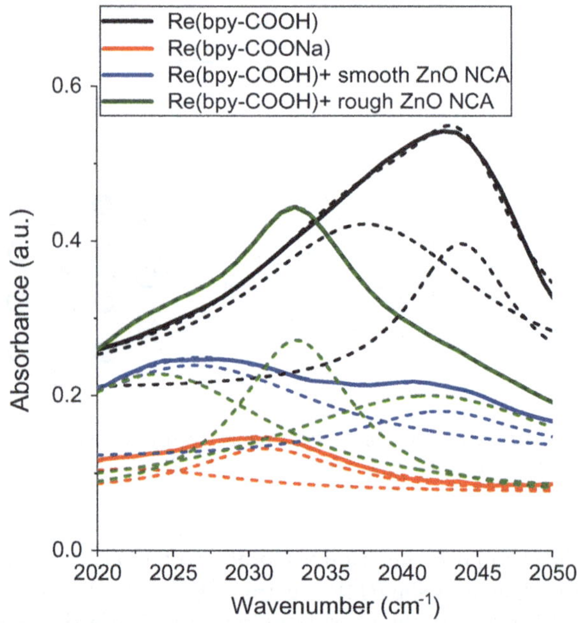

Fig. 4.25 FTIR spectra. Black: Re(bpy-COOH), red: Re(bpy-COONa), blue: smooth ZnO NCA with Re(bpy-COOH), and green: rough ZnO NCA with Re(bpy-COOH). Each spectrum was fitted by several Lorentzian peaks (dashed curves). These figures are reprinted with permission from [48], Copyright © 2019 IOP publishing

Table 4.1 Fitting parameters used in Fig. 4.26. The table is reprinted with permission from [48], Copyright © 2019 IOP publishing.

	P1 (nm)	P2 (nm)	P3 (nm)
Re(bpy-COOH)	233.4	308.8	437.0
Re(bpy-COONa)	232.3	311.1	408.9
Re(bpy-COOH)+ smooth ZnO NCA	225.5	311.8	414.9
Re(bpy-COOH)+ rough ZnO NCA	220.7	307.0	447.6

peaks around P1 (230 nm), P2 (310 nm), and P3 (400 nm) as shown in dashed lines (see Fig. 4.26a). Table 4.1 shows the fitting parameters. We found that the P3 peaks were blueshifted by 28.1 nm (Re(bpy-COONa)) and by 22.1 nm (Re(bpy-COOH) with smooth ZnO NCA) with respect to Re(bpy-COOH). The blue shift for Re(bpy-COONa) was originated from the deprotonation of the carboxyl group, which resulted in destabilization of the lowest unoccupied molecular orbital (LUMO) level of the bipyridine ligand because of the weaker electron-withdrawing group COO^-. It was found additionally, the P3 peak for the Re(bpy-COOH) with the smooth ZnO NCA (414.9 nm) was redshifted by 6.0 nm, in comparison with that for Re(bpy-COONa) (408.9 nm).

We also evaluated for the Re(bpy-COOH) with the rough ZnO NCA in a similar manner. We plotted the Kubelka-Munk plots of the diffuse reflection spectra for the

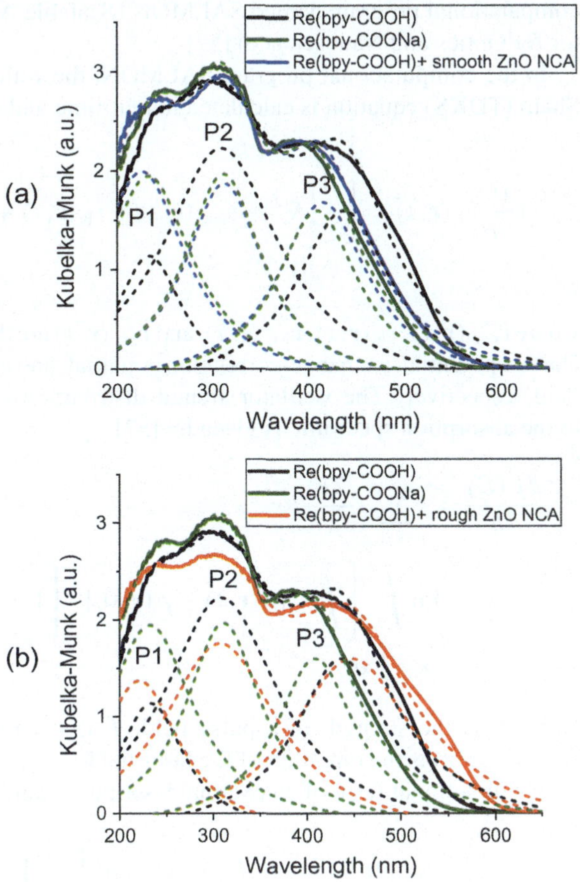

Fig. 4.26 Diffuse reflection spectra. **a** Black solid curve: Re(bpy-COOH), green solid curve: Re(bpy-COONa), blue solid curve: Re(bpy-COOH) was deposited on smooth ZnO. **b** Red solid curve: Re(bpy-COOH) was deposited on rough ZnO. These figures are reprinted with permission from [48], Copyright © 2019 IOP publishing

Re(bpy-COOH) with the rough ZnO NCA in the red curve in Fig. 4.26b. In thorough the fitting with three Lorentzian peaks, the P3 for Re(bpy-COOH) with rough ZnO NCA (447.6 nm) was redshifted by 10.6 and 38.7 nm, with respect to that of Re(bpy-COOH) and Re(bpy-COONa). These results implied that the P3 peak (~400 nm) was blueshifted by the deprotonation, i.e., by producing Re(bpy-COONa) (green curve in Fig. 4.26a, b) or by depositing onto ZnO NCA (blue curve in Fig. 4.26a and red curve in Fig. 4.26b). Furthermore, the amount of the spectral shift for longer wavelength of the Re(bpy-COOH) with rough ZnO NCA (38.7 nm: the difference of P3 peaks wavelength between Re(bpy-COOH) with rough ZnO NCA and Re(bpy-COONa)) was larger than that of smooth ZnO NCA (6.0 nm: the difference of P3 peaks wavelength between Re(bpy-COOH) with smooth ZnO NCA and Re(bpy-COONa)).

To evaluate the ONF excitation in the Re complex, we performed calculations of the photo-induced electron dynamics in the Re complex. We used a first-principles computational program named SALMON (Scalable Ab-initio Light-Matter simulator for Optics and Nanoscience) [57].

In the computational program SALMON, the following time-dependent Kohn-Sham (TDKS) equation is calculated in real-time and real-space [57]:

$$i\frac{\partial}{\partial t}\psi_j(\mathbf{r},t) = \left[-\frac{1}{2}\nabla^2 + v_{\mathrm{nuc}}(\mathbf{r},t) + v_{\mathrm{h}}(\mathbf{r},t) + v_{\mathrm{XC}}(\mathbf{r},t) + v_{\mathrm{ext}}(\mathbf{r},t)\right] \times \psi_j(\mathbf{r},t) \tag{4.7}$$

where $v_{\mathrm{nuc}}(\mathbf{r},t)$, $v_{\mathrm{h}}(\mathbf{r},t)$, $v_{\mathrm{XC}}(\mathbf{r},t)$, and $v_{\mathrm{ext}}(\mathbf{r},t)$ are the nuclear attraction potential, Hartree potential, exchange-correlation potential, and external potential of an applied field, respectively. The oscillator strength distribution $df(E)/dE$, which corresponds to the absorption spectrum, is given by [57]

$$\frac{df(E)}{dE} = \sum_{v\in x,y,z}\frac{2E}{3\pi F}\times \mathrm{Im}\int_0^T\left[\int r_v\left[\rho(\mathbf{r},t)-\rho(\mathbf{r},0)\right]\cdot\left\{1-3\left(\frac{t}{T}\right)^2+2\left(\frac{t}{T}\right)^3\right\}d\mathbf{r}\right] \times e^{iEt}dt \tag{4.8}$$

Here, F is the strength of a pulse field, T is the total propagation time, and r_v $(v \in x,\ y,\ z)$ is the direction of the laser field.

The functional form of an external laser pulse can be written as

$$v_{\mathrm{ext}}(\mathbf{r},t) = -r_v F\sin(\omega t)\sin^2\left(\frac{\pi t}{\tau}\right) \qquad (0 < t < \tau) \tag{4.9}$$

Here, ω is the laser frequency and τ determines the laser pulse duration. F is related to the laser intensity as $I = \frac{1}{2}\varepsilon_0 c F^2$, where ε_0 is the permittivity of vacuum and c is the speed of light.

In the simulation of ONF excitations, using an oscillating dipole as the source of the ONF field, the potential of the ONF, given by

$$v_{\mathrm{near}}(\mathbf{r},t) = -\frac{M\left(r_\mu - r_p\right)}{\left|\mathbf{r}-\mathbf{r}_p\right|^3}\sin(\omega t)\sin^2\left(\frac{\pi t}{\tau}\right) \tag{4.10}$$

is added to (4.7). When the ONF is utilized, the photo-absorption intensity cannot be calculated using (4.8) because of the non-uniform spatial distribution of the ONF. Thus, we evaluate the transition probability, which is defined as

$$P(t) \equiv \sum_{\mathrm{occ,unocc}}\int d\mathbf{r}\left|\varphi^*_{\mathrm{unocc}}(\mathbf{r},0)\,\varphi_{\mathrm{occ}}(\mathbf{r},t)\right|^2 \tag{4.11}$$

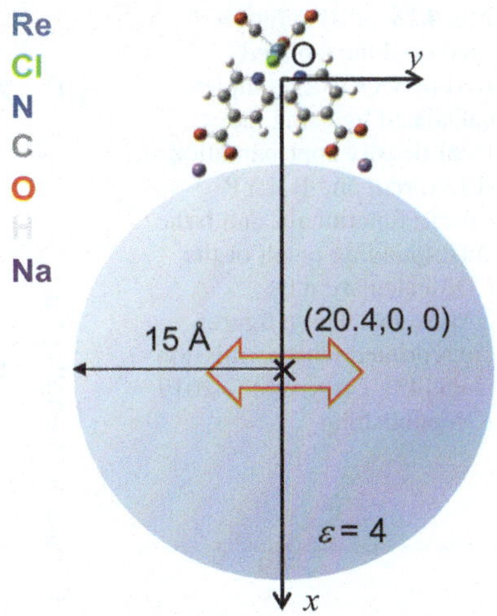

Fig. 4.27 Schematic of the simulated system. These figures are reprinted with permission from [48], Copyright © 2019 IOP publishing

where integration with respect to r is performed for the total computational space, and the sub-scripts "unocc" and "occ" denote the unoccupied and occupied molecular orbitals, respectively. Equation (4.11) can be applied to both the far-field and near-field excitations.

Figure 4.27 shows the simulated system. The geometrical optimization is performed using the LANL2DZ basis set with the Gaussian 09 program package. The optimized geometry is employed for calculations with SALMON in which photo-induced electron dynamics is simulated in real-time and real-space. The computational box size and the grid spacing are set to 30 Å× 30 Å× 30 Å and $\Delta x = \Delta y = \Delta z = 0.20$ Å, respectively. The effective core potentials (ECPs) are obtained using the Troullier-Martins scheme of the fhi98PP program [58, 59]. The real-time propagations are simulated using a constant time step of 0.8×10^3 fs. The laser intensity and duration are set to 2.8×10^{10} W/cm^2 and 30 fs, respectively. $P(t = 30$ fs) given by (4.11) is then defined as the absorption intensity.

The local-density approximation (LDA) functional is employed to calculate the exchange-correlation potential. To investigate the dependence on the functional, we perform conventional linear-response time-dependent density functional theory (TDDFT) calculations of the isolated Re complex using the B3LYP functional with the Gaussian 09 program package [60]. Figure 4.28 shows the oscillator strength calculated by the B3LYP and LDA functionals. Although the absorption energies of the LDA calculation are approximately 0.5 to 1.0 eV lower than those of the B3LYP calculation, as indicated by the colored arrows, their spectral shapes are analogous to each other. Furthermore, irrespective of the LDA or B3LYP, the absorption in a low-energy region around 2 eV is mainly attributed to the second-highest occupied molecular orbital (HOMO-1) → LUMO transition and has the MLCT characteristics. Thus, we conclude that the functional dependence does not affect, at least, the discussion of this study. It is also noted that the calculation using the B3LYP func-

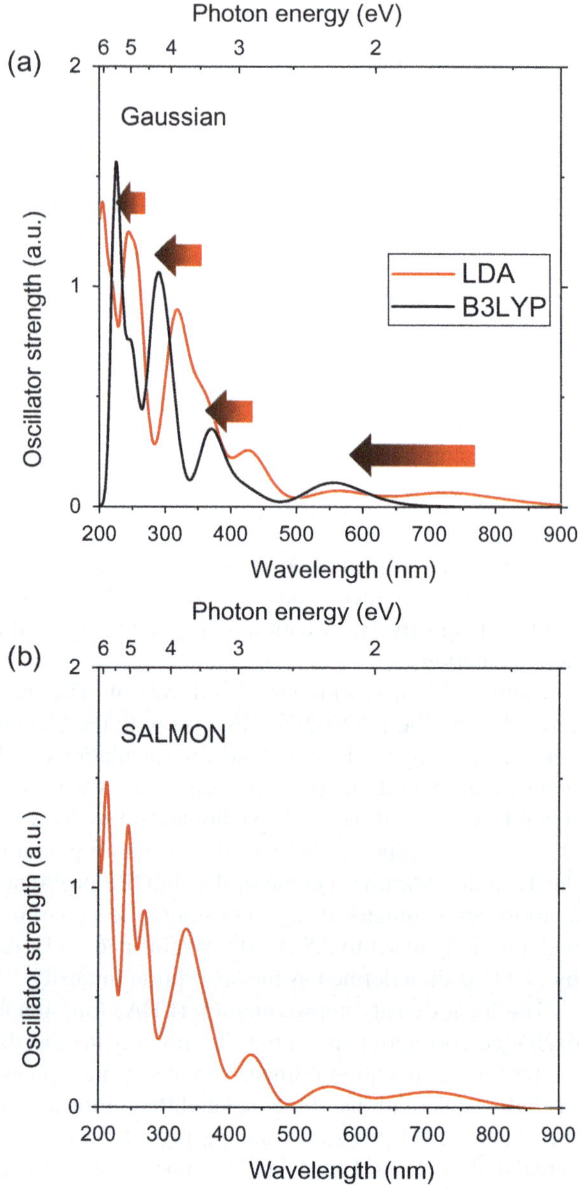

Fig. 4.28 **a** Absorption spectra of the isolated Re(bpy-COONa) complex calculated by using the local-density approximation (LDA, red) and B3LYP (black) functionals, and **b** the corresponding result of the LDA calculation by SALMON. These figures are reprinted with permission from [48], Copyright © 2019 IOP publishing

tional is time consuming when the real-time and real-space calculation is performed. Thus, we consider the detailed investigation of the DFT functional dependence to be beyond the scope of this study.

We consider the contribution of the rough ZnO surface to the photo-excitation, by utilizing a point dipole field. This is because the Re complex attached to the

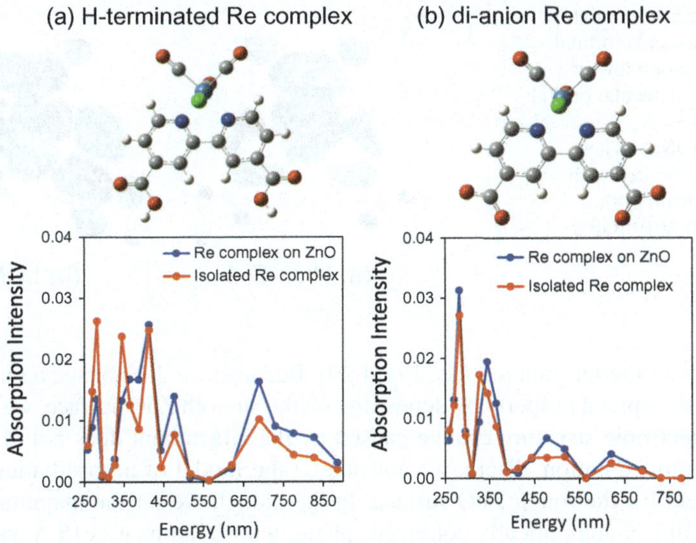

Fig. 4.29 Absorption spectra of **a** the H-terminated system and **b** the di-anion complex where the carboxyl groups are not terminated. These figures are reprinted with permission from [48], Copyright © 2019 IOP publishing

smooth ZnO surface yields an absorption spectrum analogous to that of the Re(bpy-COONa); that is, the electronic structure change in the ground state by the attachment plays a minor role in the photo-absorption. It is noted that the dipolar field is rather small within a smooth sphere whose diameter is a few hundreds of nanometers. However, if surface asperities of a particle are sufficiently smaller in size than the laser wavelength, the dipolar field can be well generated around them. We place the ONF source at the bottom of the Re complex because its carboxyl groups are bound to the ZnO surface. The dipole is separated by 16 Å from the nearest oxygen atom. The magnitude of the dipole (M in (4.10)) is then set to the same as that of a photo-polarized continuum sphere with radius (R) 15 Å and permittivity (ϵ) 4 from $M = (\epsilon - 1)/(\epsilon + 2)R^3 F$; thus, the surface of the sphere is assumed to be separated from the Re complex by 1 Å.

To evaluate the variations in the optical properties by the termination of the carboxyl groups, we calculate (a) the H-terminated system and (b) the di-anion complex where the carboxyl groups are not terminated. Their absorption spectra are shown in Fig. 4.29. The intensity increases because of the attachment to the ZnO surface when the wave number is longer than 400 nm. This intensity variation caused by the attachment is also observed in the Na-terminated Re complex.

We used the carboxyl groups with Na termination. This is because the electronic properties of Re(bpy-COONa) are similar to that of the Re complexes which deposit on ZnO aggregates. Although the absorption spectra change by the different termination, the discussion regarding the near-field effect does not affect by the spectral

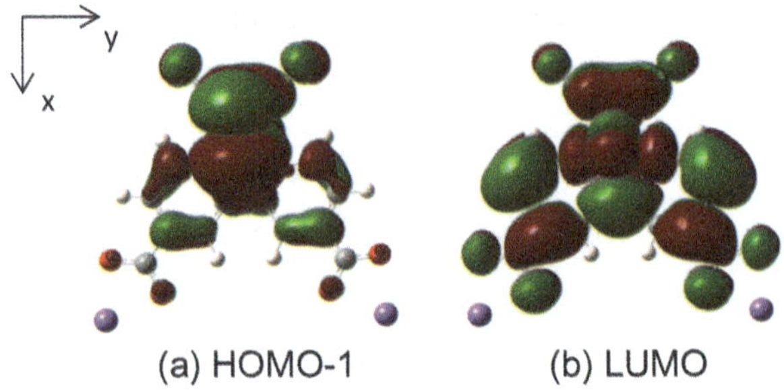

Fig. 4.30 **a** Second-highest occupied molecular orbital (HOMO-1) and **b** lowest unoccupied molecular orbital (LUMO) of Re(bpy-COONa). These figures are reprinted with permission from [48], Copyright © 2019 IOP publishing

change due to the termination (see Fig. 4.29). Because we did not see a remarkable change in the optical property by depositing on the smooth ZnO surface, we consider that the electronic structure change caused by the attachment does not play a rule for the photo-excitation. There, we considered the model of an oscillating dipolar electric field for the rough ZnO surface. In the model, we set the magnitude is the same as that from an optically polarized sphere, whose radius was 15 Å and dielectric constant (ϵ) was 4 [61]. Figure 4.30a, b is the orbitals of Re(bpy-COONa) for the second-highest occupied molecular orbital (HOMO-1) and LUMO. The energy difference between HOMO-1 and LUMO is ~1.61 eV (λ of 770 nm).

We obtained the absorption intensity (Fig. 4.31a) using all the occupied and unoccupied molecular orbitals. Although the absorption energy is underestimated due to the local-density approximation (LDA) functional in this calculation, the character of the electronic transition of the Re complex only (i.e., without ZnO) is confirmed to be almost same by using the B3LYP functional that well reproduces the absorption energy (Fig. 4.28). From the comparison of simulation and experimental results, we considered three regions in the spectra: (I) 250–300 nm, (II) 300–500 nm, and (III) 500–800 nm. These three regions correspond to the experimental peaks around 250, 300, and 450 nm. We found that the Re complex with ZnO had larger the intensity in region III, while smaller in region I, than that of the Re complex only. We also observed similar dependencies in the experimental results (Fig. 4.26b). Although the observed absorption intensity of the Re complex with ZnO decreases in region II, an intensity increase is mainly observed in the theoretical result. This incongruity in region II might be due to the photo-absorption of ZnO as shown in Fig. 4.25.

To confirm the change in the intensity due to the deposition on the rough ZnO surface, we calculated the polarization dependence of the absorption (Fig. 4.31b). It was found that independent of the wavelength, the Re complex with ZnO enhanced the absorption intensity of the x-polarization and degraded the absorption intensity of the y-polarization. Furthermore, we found that absorption peak in 400–800 nm was originated by the x-polarization. While, the absorption peak around 300 nm was originated by the y-polarization. These results indicated that the polarization dependence of the absorption spectra depended on the wavelength. The dependence of the absorption intensity on the incident polarization should be reasonable from the electronic transition character. As shown in Fig. 4.30a, b, HOMO-1 consisted

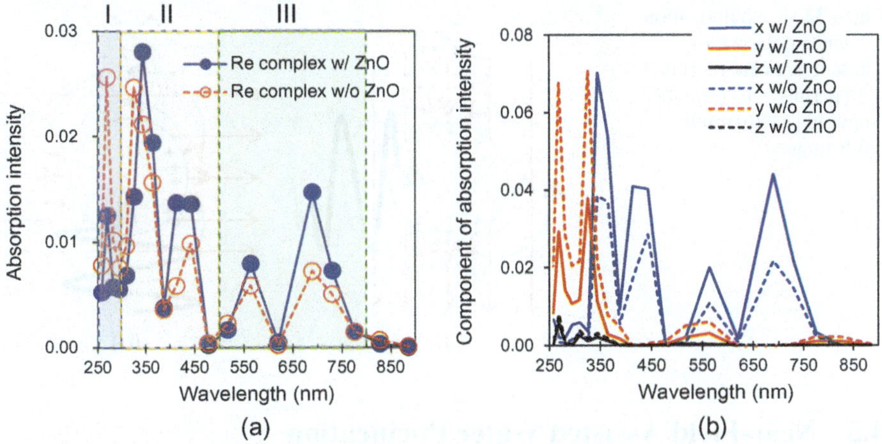

Fig. 4.31 **a** Photo-absorption intensity spectra of the Re complex only (w/o ZnO, red open circles and dashed line) and the Re complex deposited on ZnO (w/ ZnO, blue solid circles and line). **b** Polarization (x, y, and z) dependence of absorption intensity of the Re complex with ZnO (w/ ZnO, solid lines) and without ZnO (w/o ZnO, dashed lines). These figures are reprinted with permission from [48], Copyright © 2019 IOP publishing

of the Re d-orbital, however, LUMO had a large distribution around the aromatic ligand moiety. From these results, the absorption in region III was originated from the MLCT excitation as well as the ethyl-terminated system [54]. Because the HOMO-1 → LUMO transition moment has a large x-component, the absorption is strongly induced by the x-polarization. As for $\lambda < 400$ nm, an electronic transition is induced between the ligand moieties, which causes a large transition moment along the y-direction.

Figure 4.32 illustrates the polarization dependence of the absorption intensity owing to the deposition on the rough ZnO surface. In the case of x-polarization (Fig. 4.32a), its direction has the same as the dipole field induced by the electronic polarization of ZnO aggregates (as indicated by the red dashed circle). Therefore, the electric field intensity increases, and it results in the absorption intensity of Re complex. In contrast, the directions of the y and z-polarization are antiparallel with respect to the dipole field direction, and it results in cancel themselves (blue dashed circle in Fig. 4.32b); as a result, the deposition of Re complex on the rough ZnO surface results in decrease in the absorption intensity. Our numerical calculations revealed the atomic-level details of the absorption intensity owing to the deposition on the rough ZnO surface.

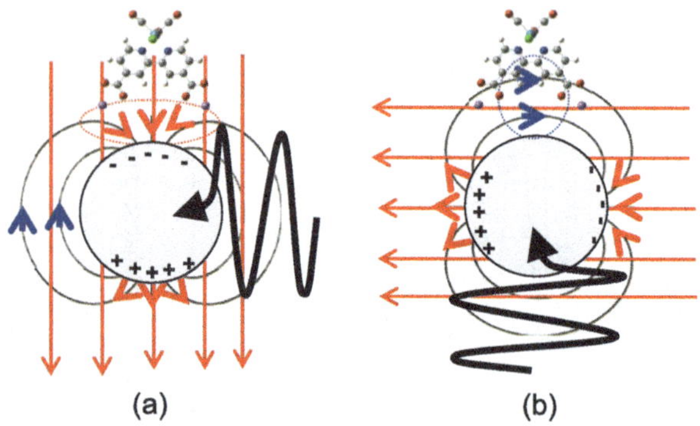

Fig. 4.32 **a** *x* polarization. **b** *y* and *z* polarization. These figures are reprinted with permission from [48], Copyright © 2019 IOP publishing

4.5 Near-Field Assisted Water Purification

Deoxyribonucleic acid (DNA) strands consist of four types of bases, including adenine (A), guanine (G), cytosine (C), and thymine (T). The genetic information is determined by their sequence. The development of artificial techniques for controlling several of the molecular reactions of DNA strands is a fundamental issue for experimental study on artificial protein synthesis, regional genetic curing, and the stochastic analysis of several genetic expressions [62, 63]. The photo-excitation technique for controlling molecular reactions is one of the most useful techniques because it is easy to access with various parameters including wavelength, polarization, light intensity [64]. Unlike higher LET radiations, such as x-rays and γ-rays, DNA strands reveal only a few absorbances on irradiation of ultraviolet (UV), visible, and infrared (IR) lights. On the other hand, DNA strands demonstrate particularly high absorbance at specific wavelengths in the deep UV (DUV) region of light with more than 4.0 eV of optical energy, corresponding to their base sequence and corresponding binding structures. Therefore, the light irradiation to DNA strands results in degradation and reconstruction of the binding structure as well as biochemical properties. In addition, combination with some photo-responsive organic molecules, including stilbenes [65], diarylethenes [66], spiropyrans [67], and azobenzenes [68] has been investigated to realize variable control of molecular reactions. While such methods with light irradiation have quite valuable characteristics, excessive DUV light irradiation necessarily induces some unexpected reactions, such as the damaging and breaking of DNA strands due to electronic dissociation [69].

Here, we focus on thymine dimerization by non-uniform ONF [70]. Thymine dimerization [71] is a major photo-chemically induced destruction. Two thymine (T) sitting next to each other in a DNA strand, i.e., TT doublet, are preferably dimerized as a thymine dimer by irradiating DUV light (Fig. 4.33). The thymine dimers formation results in messenger RNA transcription errors, leading to a complex web of biological responses such as apoptosis, immune suppression, and carcinogenesis [72–74].

The dimerization process is originated from electronic excitation and the corresponding orientation of the reacting double bonds in each molecule [71]. To realize

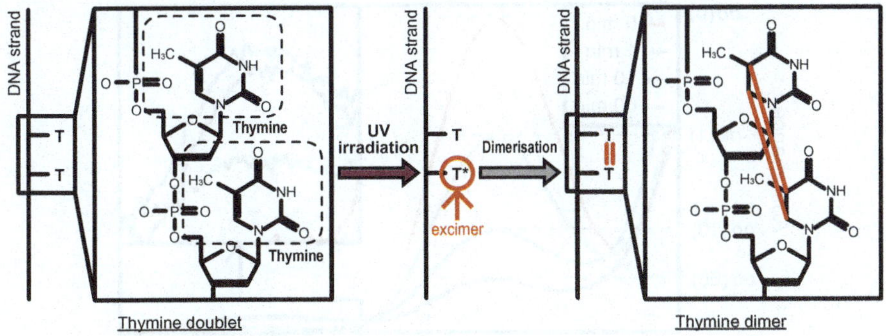

Fig. 4.33 Schematic of the photo-dynamics of thymine dimerization. The figure is reprinted from [70] under the terms of the Creative Commons Attribution license

dimerization, it is required to excite the $\pi - \pi^*$ transition (~4.7 eV), and the excited T is referred to as T* (see Fig. 4.33). Then, some contents of excited states can decay and form dimer products as referred to as TT. Therefore, propagating FF light should have higher photon energy than 4.7 eV for conventional optical dimerization. Here, we investigated a new photo-dimerization via vibrational excitation using non-uniform ONF. As discussed in the Chap. 2, in the case that ONF is affected, we expect to excite the intermediate vibrational levels, and resultant multi-step excitation. In that case, we can use shorter wavelength than 300 nm to maximally inhibit damage to RNA strands.

Before the demonstration of the dimerization using non-uniform ONF, we verified the absorption peak decreased of the poly-T molecule by irradiating DUV light. We used laser diodes (LD) with $\lambda = 266$ nm as a conventional DUV light source with 3 mW. As with the following experiments, in order to minimize temperature elevation due to photo-thermal effect, a solution is set in ice water during laser irradiation. Figure 4.34 shows the change of absorption spectra during the irradiation and the decrement of the peak absorption at a wavelength of 267 nm, respectively.

As shown, the absorption peak at λ 267 nm gradually decreased during the irradiation, and the dimerization procedure was indirectly monitored by the decrement. Furthermore, the absorption peak behavior at a wavelength of around 325 nm was also observed as above as the peak value incremented until reaching 10 min of irradiation. Origin of the peak has been known as formation of (6-4) photo-products [75], which is obtained as a result of the bonding of a TT doublet by a single bond. Another important finding of the result is that exceeding 60 min of DUV light irradiation resulted in degradation of the poly-T strands, which was confirmed by a decrement of the absorption intensity.

To clarify the spectral change, we defined the amount of the decrement as $-\Delta\alpha d(t)$, which is calculated from the peak values of absorbance before and after t-min irradiation, $\alpha d(0)$ and $\alpha d(t)$, respectively. Then, the absorbed optical energy $E_{\text{abs}}(t)$ during the light irradiation is defined as $E_{\text{abs}}(t) = (1\text{–}10^{-\Delta\alpha d(t)})It$, where I

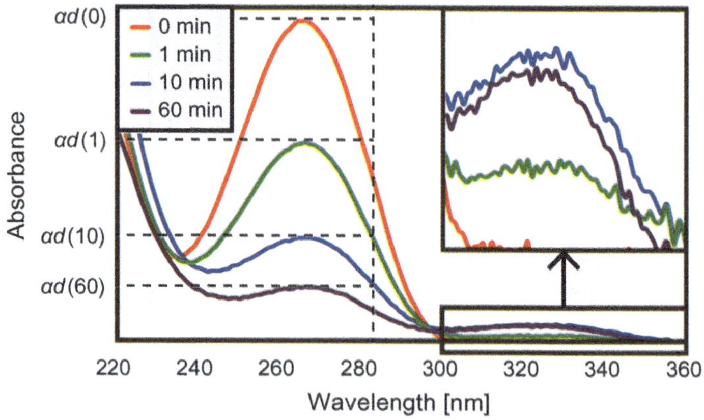

Fig. 4.34 Decrement and increment of absorption peak values due to product of thymine dimers and (6-4) photo-products, respectively, by irradiating DUV lights. The figure is reprinted from [70] under the terms of the Creative Commons Attribution license

Fig. 4.35 Relation between amount of decrement $\Delta \alpha d(t)$ and absorbed optical energy $E_{\text{abs}}(t)$ depending on irradiation of light with a wavelength of 266 nm. The figure is reprinted from [70] under the terms of the Creative Commons Attribution license

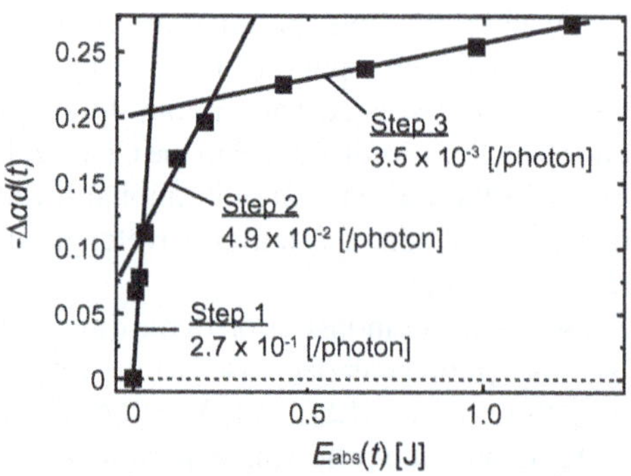

is the power of the irradiated light. The resulting absorption peak decrease shown in Fig. 4.34, $-\Delta \alpha d(t)$, were plotted as a function of $E_{\text{abs}}(t)$ (Fig. 4.35).

As can be seen in Fig. 4.35, the decrement amount of the absorbance can be approximated by three fitting lines with gradients of 2.7×10^{-1}/photon, 4.9×10^{-2}/photon, and 3.5×10^{-3}/photon, respectively. Each gradient corresponds to the quantum efficiency of each molecular reaction. Amounts of $E_{\text{abs}}(t)$ on the third line correspond to $60 \sim 240$ min of irradiation, in which the poly-T strands are destructed as shown in Fig. 4.34. Namely, the each reactions of dimerization can be considered as follows:

Step 1. Random dimerization between arbitral TT doublets in poly-T strands occurs.

Step 2. Dimerization is suppressed owing to the decrease in the number of remaining TT doublets and (6-4) photo-products are produced.

Step 3. Dimerization is saturated and poly-T strands are damaged and destroyed due to excess absorption of photon.

Reacting yields owing to photo-excitation were mainly confirmed by the changes of absorbance. The absorbance of DNA strands is mainly exhibited in the DUV region, which is greater than 3.2 eV. Combinations of four types of bases that are used to construct DNA strands and the related bonding energies between each base determine such energies. Thus, degradation and reconstruction of their bonding structures during particular molecule reactions necessarily affect changes of their absorption peaks.

As can be seen in Fig. 4.33, cleavages of double bonding in T bases and resultant covalent bonding between the two occurred by dimerization. In the process, the value of absorbance is gradually changed owing to the bonding structure of T bases. Namely, the absorption peak value at $\lambda \sim 267$ nm, which relates to the existence of double bonding within the pyrimidine ring of T, is evidently decreased by the dimerization. Furthermore, (6-4) photo-products are additionally obtained as by-products of the dimerization [75]. The absorption peak value is revealed at a wavelength of around 325 nm, differing from dimerization. Finally, by additional irradiation of light, all strands of the thymine dimers and the (6-4) photo-products are degraded, and their absorption peaks are decremented.

In experiments, we utilized the decrement of the peak value as a quantitative evaluation indicator of the dimerization. In addition, commercially available poly-T strands, which consist of 60 T bases with 20 nm length, were prepared to maximize the decrement owing to the dimerization. Each poly-T strand was terminated by thiol. We used 200 μL of solution, in which the final density of poly-T strands was 0.32 μM. By using such diluted solutions, saturation of absorbance can be avoided.

To realize the NF excitation, the source of the ONF must be generated in the vicinity of poly-T strands. To meet this requirement, as can be seen in Fig. 4.36, poly-T strands were directly attached to an Au nanoparticle by Au-thiol bonding, which is known as a quite stable bonding under the irradiation of light. Here, an Au nanoparticle generated the ONF. We used Au nanoparticles with the average diameter of 20 nm. The size of spatial distribution of ONF is considered to be comparably to the size of nanostructure [76]; thus, it is expected the ONF covered the whole 60 T bases of poly-T strands. In the experiment, the number of bonded poly-T strands to the Au nanoparticle was assumed exceeding 500. To realize the affective bonding process, we mixed 5.0 nM of Au nanoparticles and 2.5 μM of poly-T strands. Next, we leaved the mixture at room temperature for 24 h. Then, we diluted the mixture by adding distilled water and 200 μL of the sample solution. The mixing condition was confirmed by results of electrophoresis which migrates un-bonded poly-T strands.

For experimental demonstration of dimerization with NF effect, LDs with λ of 325 and 405 nm irradiated sample solutions. In addition to the sample contained poly-T bonded-Au nanoparticles, we prepared the sample contained only poly-T strands as a reference. We performed the similar procedure to monitor changes of absorptions as in the previous DUV light irradiation. Respective output intensities of LDs were 25 mW (325 nm) and 100 mW (405 nm). Figure 4.37 is the changes of the decrement by LDs irradiation. To exclude the effect of the absorbance due to Au nanoparticles, we processed differential absorption spectroscopy instead of the absorbance from the measured result.

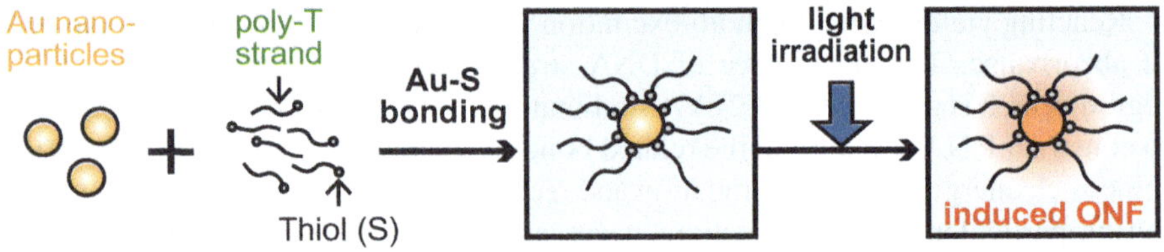

Fig. 4.36 Schematic diagram of bonding poly-T strands to Au nanoparticles as source of the ONF and excitation of the ONF by irradiation of light. The figure is reprinted from [70] under the terms of the Creative Commons Attribution license

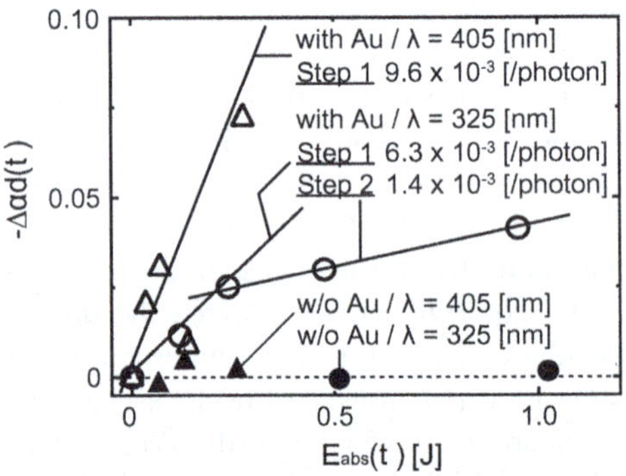

Fig. 4.37 Relation between $-\Delta\alpha d(t)$ and $E_{\mathrm{abs}}(t)$ depending on the irradiation of 325 and 425 nm wavelength light. The figure is reprinted from [70] under the terms of the Creative Commons Attribution license

As can be seen in Fig. 4.37, only the case of poly-T strands with Au nanoparticles (open triangles and circles) decreased in peak absorptions. Furthermore, we found that the results were fitted using two lines, not three as can be seen in Fig. 4.35 with similar amounts of $E_{\mathrm{abs}}(t)$ with DUV excitation.

The facts indicate that there is no direct effect from irradiation or molecular reactions by visible light excitation. However, in the case of bonding with Au nanoparticles, it is reasonable to say that the process can be occurred by the generation of ONF with 325 and 405 nm light irradiation, as in the case of irradiating DUV light. Because the process by 325 and 405 nm light irradiation has only two steps and does not include Step 3 as shown in Fig. 4.35, we consider that the process does not involve any damaging and destruction of poly-T strands. As one of the evidences of our postulate, Fig. 4.38 indicates absorbance of poly-T strands by 325 nm light irradiation.

As shown, there is no clear peak at around a wavelength of 325 nm regardless of the strands being bonded to Au nanoparticles. This means that 240 min of irradiation of visible light cannot induce the forming of (6-4) photo-products and the following damaging and breaking of strands.

The difference of quantum efficiencies between the cases of irradiation with wavelengths of 325 and 405 nm is discussed as due to the differences between the absorption edge wavelength of the Au nanoparticle. That is, while the absorption edge wavelength is obtained at around 300 nm, wavelength 325 nm is closer to the absorp-

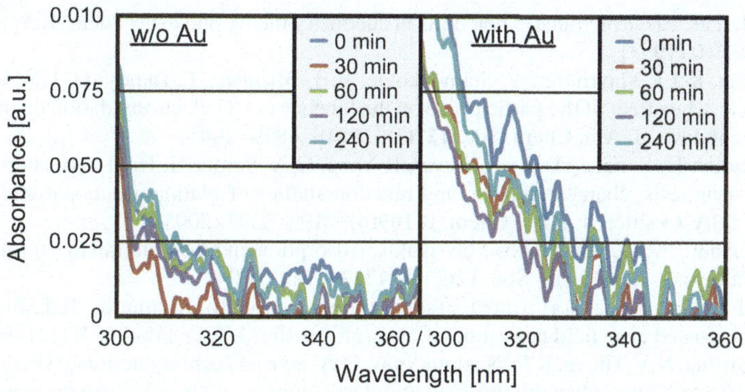

Fig. 4.38 Absorbance of poly-T strands by 325 nm light irradiation. Right panel: with nanoparticles. Left pane: without Au nanoparticles. The figure is reprinted from [70] under the terms of the Creative Commons Attribution license

tion edge than 405 nm, so that larger amounts of optical energy are involved in the interaction between Au material and larger ONF are preferably occurred in the case of irradiation of wavelength 325 nm of light. For this reason, the case of irradiation of light at a 325 nm wavelength reached Step 2, revealing much faster convergence than the other.

We performed the photo-excitation of DNA by introducing non-uniform ONF with lower photon energy than the that of absorption band edge. In our experiment, we used Au nanoparticles as the source of the ONF to verify the dimerization of poly-T strands. The results show an evident amount of dimerization, which was indicated by a change of the absorption spectrum and was obtained only in the case of the strands with nanoparticles. These results supported that ONF could realize multi-step excitation of π-electron and subsequent dimerization. In addition, the result indicates that any damaging and breaking in poly-T strands cannot occur by applying similar amounts of optical power as irradiation by DUV light.

References

1. A. Fujishima, K. Honda, Photolysis-decomposition of water at the surface of an irradiated semiconductor. Nature **238**, 37 (1972)
2. T. Inoue, A. Fujishima, S. Konishi, K. Honda, Photoelectrocatalytic reduction of carbon dioxide in aqueous suspensions of semiconductor powders. Nature **277**, 637–638 (1979)
3. Kazuhiko Maeda, Kentaro Teramura, Lu Daling, Tsuyoshi Takata, Nobuo Saito, Yasunobu Inoue, Kazunari Domen, Photocatalyst releasing hydrogen from water. Nature **440**, 295 (2006)
4. H. Takeda, O. Ishitani, Development of efficient photocatalytic systems for CO_2 reduction using mononuclear and multinuclear metal complexes based on mechanistic studies. Coord. Chem. Rev. **254**(3), 346–354 (2010). Inorganic Reaction Mechanisms
5. P. Atkins, J. De Paula, J. Keeler, *Atkins' Physical Chemistry* (Oxford University Press, Oxford, 2017)

6. R.A. Marcus, Electron transfer reactions in chemistry theory and experiment. Rev. Mod. Phys. **65**, 599–610 (1993)
7. M. Arenz, K.J.J. Mayrhofer, V. Stamenkovic, B.B. Blizanac, T. Tomoyuki, P.N. Ross, N.M. Markovic, The effect of the particle size on the kinetics of CO electrooxidation on high surface area Pt catalysts. J. Am. Chem. Soc. **127**(18), 6819–6829 (2005)
8. R.M. Rioux, H. Song, J.D. Hoefelmeyer, P. Yang, G.A. Somorjai, High-surface-area catalyst design: synthesis, characterization, and reaction studies of platinum nanoparticles in mesoporous SBA-15 silica. J. Phys. Chem. B **109**(6), 2192–2202 (2005)
9. Xu Ye, Andrei V. Ruban, Manos Mavrikakis, Adsorption and dissociation of O_2 on Pt-Co and Pt−Fe alloys. J. Am. Chem. Soc. **126**(14), 4717–4725 (2004)
10. Takashi Yatsui, Kazaunori Iijima, Tsubasa Imoto, Kokoro Kitamura, Tadashi Kawazoe, Phonon-assisted near-field activation of electron transfer. J. Nanophoton. **7**(1), 1–8 (2013)
11. G.A. Tsirlina, N.V. Titova, R.R. Nazmutdinov, O.A. Petrii, Electroreduction of $[Fe(CN)_6]^{3-}$ on a mercury electrode: substantiating activationless character of the process at high overvoltages. Russ. J. Electrochem. **37**, 15–25 (2001)
12. M.H. Chakrabarti, E.P. Roberts, Analysis of mixtures of ferrocyanide and ferricyanide using uv-visible spectroscopy for characterisation of a novel redox flow battery. J. Chem. Soc. Pak. **30**(6), 817–823 (2008)
13. B. Yazici, Hydrogen evolution at platinum (Pt) and at platinized platinum (Ptz) cathodes. Turk. J. Chem. **23**(6), 301–308 (1999)
14. Lakshmi Krishnan, Steven E. Morris, Glenn A. Eisman, Platinum black polymer electrolyte membrane based electrodes revisited. J. Electrochem. Soc. **155**(9), B869–B876 (2008)
15. Z. Zou, J. Ye, K. Sayama, H. Arakawa, Direct splitting of water under visible light irradiation with an oxide semiconductor photocatalyst. Nature **414**, 625–627 (2001)
16. X. Wang, K. Maeda, A. Thomas, K. Takanabe, G. Xin, J.M. Carlsson, K. Domen, M. Antonietti, A metal-free polymeric photocatalyst for hydrogen production from water under visible light. Nat. Mater. **8**(1), 76–80 (2009)
17. Takashi Yatsui, Tsubasa Imoto, Takahiro Mochizuki, Kokoro Kitamura, Tadashi Kawazoe, Dressed-photon-phonon (DPP)-assisted visible- and infrared-light water splitting. Sci. Rep. **4**, 4561 (2015)
18. W.I. Park, D.H. Kim, S.-W. Jung, G-C. Yi, Metalorganic vapor-phase epitaxial growth of vertically well-aligned ZnO nanorods. Appl. Phys. Lett. **80**(22), 4232–4234 (2002)
19. K. Kitamura, T. Yatsui, M. Ohtsu, G.-C. Yi, Fabrication of vertically aligned ultrafine ZnO nanorods using metal–organic vapor phase epitaxy with a two-temperature growth method. Nanotechnology **19**(17), 175305 (2008)
20. Kokoro Kitamura, Takashi Yatsui, Motoichi Ohtsu, Observation of quantum confinement in Zno nanorods fabricated using a two-temperature growth method. Appl. Phys. B **97**(4), 825–828 (2019)
21. Manfred Mascheck, Stephen Schmidt, Martin Silies, Takashi Yatsui, Kokoro Kitamura, Motoichi Ohtsu, David Leipold, Erich Runge, Christoph Lienau, Observing the localization of light in space and time by ultrafast second-harmonic microscopy. Nat. Photon. **6**(5), 283–292 (2012)
22. S. Baruah, S.S. Sinha, B. Ghosh, S.K. Pal, A.K. Raychaudhuri, J. Dutta, Photoreactivity of ZnO nanoparticles in visible light: Effect of surface states on electron transfer reaction. J. Appl. Phys. **105**(7), 074308 (2009)
23. S. Baruah, M.A. Mahmood, M.T.Z. Myint, T. Bora, J. Dutta, Enhanced visible light photocatalysis through fast crystallization of zinc oxide nanorods. Beilstein J. Nanotechnol. **105**(1), 14–20 (2010)
24. W.I. Park, G.-C. Yi, M.Y. Kim, S.J. Pennycook, Quantum confinement observed in ZnO/ZnMgO nanorod heterostructures. Adv. Mater. **15**(6), 526–529 (2003)
25. K. Vanheusden, C.H. Seager, W.L. Warren, D.R. Tallant, J.A. Voigt, Correlation between photoluminescence and oxygen vacancies in zno phosphors. Appl. Phys. Lett. **68**(3), 403–405 (1996)
26. D.C. Reynolds, D.C. Look, B. Jogai, H. Morkoç, Similarities in the bandedge and deep-centre photoluminescence mechanisms of Zno and gan. Solid State Commun. **101**(9), 643–646 (1997)

27. D.C. Reynolds, D.C. Look, B. Jogai, Fine structure on the green band in Zno. J. Appl. Phys. **89**(11), 6189–6191 (2001)
28. Allen J. Bard, Mark S. Wrighton, Thermodynamic potential for the anodic dissolution of n-type semiconductors. J. Electrochem. Soc. **124**(11), 1706–1710 (1977)
29. T. Freund, S.R. Morrison, Mechanism of cathodic processes on the semiconductor zinc oxide. Surf. Sci. **9**(1), 119–132 (1968)
30. K. Gelderman, L. Lee, S.W. Donne, Flat-band potential of a semiconductor: using the Mott-Schottky equation. J. Chem. Educ. **84**(4), 685 (2007)
31. Thu Hac Huong Le, Kazuma Mawatari, Yuriy Pihosh, Tadashi Kawazoe, Takashi Yatsui, Motoichi Ohtsu, Masahiro Tosa, Takehiko Kitamori, Optical near-field induced visible response photoelectrochemical water splitting on nanorod tio_2. Appl. Phys. Lett. **99**(21), 213105 (2011)
32. K. Robbie, M.J. Brett, Sculptured thin films and glancing angle deposition: growth mechanics and applications. J. Vac. Sci. Technol. A **15**(3), 1460–1465 (1997)
33. I. Turkevych, Y. Pihosh, M. Goto, A. Kasahara, M. Tosa, S. Kato, K. Takehana, T. Takamasu, G. Kido, N. Koguchi, Photocatalytic properties of titanium dioxide sputtered on a nanostructured substrate. Thin Solid Films **516**(9), 2387–2391 (2008). The 7th International Conference on Nano-Molecular Electronics (ICNME 2006)
34. M.A.A. Clyne, B.A. Thrush, Mechanism of chemiluminescent combination reactions involving oxygen atoms. Proc. R. Soc. Lond. A **269**(1338), 404–418 (1962)
35. M.P. Srivastava, A. Kobayashi, Carbon dioxide decomposition by plasma methods and application of high energy and high density plasmas in material processing and Nanostructures. Trans. JWRI **39**(1), 11–25 (2010)
36. Y. Tamaura, K. Nishizawa, CO_2 decomposition into C and conversion into CH_4 using the H_2-reduced magnetite. Energy Convers. Manag. **33**(5–8), 573–577 (1992)
37. K. Teramura, T. Tanaka, H. Ishikawa, Y. Kohno, T. Funabiki, Photocatalytic reduction of CO_2 to CO in the presence of H_2 or CH_4 as a reductant over MgO. J. Phys. Chem. B **108**(1), 346–354 (2004)
38. Nabila Tanjeem, Tadashi Kawazoe, Takashi Yatsui, CO_2 phonon mode renormalization using phonon-assisted energy up-conversion. Sci. Rep. **3**, 3341 (2013)
39. http://hitran.org/
40. T.G. Slanger, G. Black, The CO_2 photolysis problem. J. Chem. Phys. **54**(5), 1889–1898 (1971)
41. T.G. Slanger, G. Black, CO_2 photolysis revisited. J. Chem. Phys. **68**(4), 1844 (1978)
42. R.P. Wayne, Photodissociation dynamics and atmospheric chemistry. J. Geophys. Res. **98**(E7), 13119–13136 (1978)
43. E.C.Y. Inn, J.M. Heimerl, The photolysis of CO_2 at wavelengths exceeding 1740 Å. J. Atmos. Sci. **28**(6), 838–841 (1971)
44. W.B. DeMore, M. Mosesman, Photolysis of CO_2 at 1849 Å. J. Atmos. Sci. **28**(6), 842–846 (1971)
45. Jeannot Hawecker, Jean-Marie Lehn, Raymond Ziessel, Photochemical and electrochemical reduction of carbon dioxide to carbon monoxide mediated by (2,2'-bipyridine)tricarbonylchlororhenium(I) and related complexes as homogeneous catalysts. Helv. Chim. Acta **69**(8), 1990–2012 (1986)
46. C.A. Craig, L.O. Spreer, J.W. Otvos, M. Calvin, Photochemical reduction of carbon dioxide using nickel tetraazamacrocycles. J. Phys. Chem. **94**(20), 7957–7960 (1990)
47. Yu. Hiroyuki Takeda, Haruki Sugiyama Monma, Hidehiro Uekusa, Osamu Ishitani, Development of visible-light driven Cu(I) complex photosensitizers for photocatalytic CO_2 reduction. Front. Chem. **7**, 418 (2019)
48. T. Yatsui, Y. Nakahira, Y. Nakamura, T. Morimoto, Y. Kato, M. Yamamoto, T. Yoshida, K. Iida, K. Nobusada, Realization of red shift of absorption spectra using optical near-field effect. Nanotechnology **30**(34), 34LT02 (2019)
49. Y. Ono, J. Nakamura, M. Hayashi, K.I. Takahashi, Effect of substituent groups in rhenium bipyridine complexes on photocatalytic CO_2 reduction. Am. J. Appl. Chem. **2**(5), 74–79 (2014)
50. Qifeng Zhang, Tammy P. Chou, Bryan Russo, Samson A. Jenekhe, Guozhong Cao, Aggregation of ZnO nanocrystallites for high conversion efficiency in dye-sensitized solar cells. Angew. Chem. Int. Ed. **47**(13), 2402–2406 (2008)

51. Qifeng Zhang, Tammy P. Chou, Bryan Russo, Samson A. Jenekhe, Guozhong Cao, Polydisperse aggregates of ZnO nanocrystallites: a method for energy-conversion-efficiency enhancement in dye-sensitized solar cells. Adv. Funct. Mater. **18**(11), 1654–1660 (2008)
52. Takashi Yatsui, Yang Ryu, Tetsu Morishima, Wataru Nomura, Tadashi Kawazoe, Tetsu Yonezawa, Masao Washizu, Hiroyuki Fujita, Motoichi Ohtsu, Self-assembly method of linearly aligning Zno quantum dots for a nanophotonic signal transmission device. Appl. Phys. Lett. **96**(13), 133106 (2010)
53. Yang Liu, Tetsu Morishima, Takashi Yatsui, Tadashi Kawazoe, Motoichi Ohtsu, Size control of sol-gel-synthesized zno quantum dots using photo-induced desorption. Nanotechnology **22**(21), 215605 (2011)
54. Jonathan M. Smieja, Clifford P. Kubiak, Re(bipy-tBu)$(CO)_3$Cl-improved catalytic activity for reduction of carbon dioxide: Ir-spectroelectrochemical and mechanistic studies. Inorganic Chem. **49**(20), 9283–9289 (2010)
55. D.A. Kurtz, K.R. Brereton, K.P. Ruoff, H.M. Tang, G.A.N. Felton, A.J.M. Miller, J.L. Dempsey, Bathochromic shifts in rhenium carbonyl dyes induced through destabilization of occupied orbitals. Inorganic Chem. **57**(9), 5389–5399 (2018)
56. P. Kubelka, New contributions to the optics of intensely light-scattering materials. part i. J. Opt. Soc. Am. **38**(5), 448–457 (1948)
57. Masashi Noda, Shunsuke A. Sato, Yuta Hirokawa, Mitsuharu Uemoto, Takashi Takeuchi, Shunsuke Yamada, Atsushi Yamada, Yasushi Shinohara, Maiku Yamaguchi, Kenji Iida, Isabella Floss, Tomohito Otobe, Kyung-Min Lee, Kazuya Ishimura, Taisuke Boku, George F. Bertsch, Katsuyuki Nobusada, Kazuhiro Yabana, SALMON: scalable ab-initio light-matter simulator for optics and nanoscience. Comput. Phys. Commun. **235**, 356–365 (2019)
58. N. Troullier, J.L. Martins, Efficient pseudopotentials for plane-wave calculations. Phys. Rev. B **43**, 1993–2006 (1991)
59. Martin Fuchs, Matthias Scheffler, Ab initio pseudopotentials for electronic structure calculations of poly-atomic systems using density-functional theory. Comput. Phys. Commun. **119**(1), 67–98 (1999)
60. M.J. Frisch, G.W. Trucks, H.B. Schlegel, G.E. Scuseria, M.A. Robb, J.R. Cheeseman, G. Scalmani, V. Barone, G.A. Petersson, H. Nakatsuji, X. Li, M. Caricato, A. Marenich, J. Bloino, B.G. Janesko, R. Gomperts, B. Mennucci, H.P. Hratchian, J.V. Ortiz, A.F. Izmaylov, J.L. Sonnenberg, D. Williams-Young, F. Ding, F. Lipparini, F. Egidi, J. Goings, B. Peng, A. Petrone, T. Henderson, D. Ranasinghe, V.G. Zakrzewski, J. Gao, N. Rega, G. Zheng, W. Liang, M. Hada, M. Ehara, K. Toyota, R. Fukuda, J. Hasegawa, M. Ishida, T. Nakajima, Y. Honda, O. Kitao, H. Nakai, T. Vreven, K. Throssell, J.A. Montgomery Jr., J.E. Peralta, F. Ogliaro, M. Bearpark, J.J. Heyd, E. Brothers, K.N. Kudin, V.N. Staroverov, T. Keith, R. Kobayashi, J. Normand, K. Raghavachari, A. Rendell, J.C. Burant, S.S. Iyengar, J. Tomasi, M. Cossi, J.M. Millam, M. Klene, C. Adamo, R. Cammi, J.W. Ochterski, R.L. Martin, K. Morokuma, O. Farkas, J.B. Foresman, D.J. Fox, Gaussian 09, revision A.02 (Gaussian, Inc., Wallingford CT, 2016)
61. Ü. Özgür, Ya I. Alivov, C. Liu, A. Teke, M.A. Reshchikov, S. Doğan, V. Avrutin, S.-J. Cho, H. Morkoç, A comprehensive review of Zno materials and devices. J. Appl. Phys. **98**(4), 041301 (2005)
62. Günter Mayer, Alexander Heckel, Biologically active molecules with a "light switch". Angew. Chem. Int. Edit. **45**(30), 4900–4921 (2006)
63. Pau Gorostiza, EhudY Isacoff, Optical switches for remote and noninvasive control of cell signaling. Science **322**(5900), 395–399 (2008)
64. Anouk S. Lubbe, Wiktor Szymanski, Ben L. Feringa, Recent developments in reversible photoregulation of oligonucleotide structure and function. Chem. Soc. Rev. **46**, 1052–1079 (2017)
65. Heiko K. Cammenga, Vladimir N. Emel'yanenko, Sergey P. Verevkin, Re-investigation and data assessment of the isomerization and 2,2'-cyclization of stilbenes and azobenzenes. Ind. Eng. Chem. Res. **48**(22), 10120–10128 (2009)
66. Sylvain L. Gilat, Stephen H. Kawai, Jean-Marie Lehn, Light-triggered molecular devices: photochemical switching of optical and electrochemical properties in molecular wire type diarylethene species. Chem. Eur. J. **1**(5), 275–284 (1995)

67. John B. Flannery, Photo- and thermochromic transients from substituted 1',3',3'-trimethylindolinobenzospiropyrans. J. Am. Chem. Soc. **90**(21), 5660–5671 (1968)
68. Anna Bergen, Sergii Rudiuk, Mathieu Morel, Thomas Le Saux, Heiko Ihmels, Damien Baigl, Photodependent melting of unmodified DNA using a photosensitive intercalator: a new and generic tool for photoreversible assembly of DNA nanostructures at constant temperature. Nano Lett. **16**(1), 773–780 (2016)
69. Badia Boudaïffa, Pierre Cloutier, Darel Hunting, Michael A. Huels, Léon Sanche, Resonant formation of DNA strand breaks by low-energy (3 to 20 ev) electrons. Science **287**(5458), 1658–1660 (2000)
70. Naoya Tate, Takashi Yatsui, Visible light-induced thymine dimerisation based on large localised field gradient by non-uniform optical near-field. Sci. Rep. **9**, 18383 (2019)
71. Wolfgang J. Schreier, Tobias E. Schrader, Florian O. Koller, Peter Gilch, Carlos E. Crespo-Hernández, Vijay N. Swaminathan, Thomas Carell, Wolfgang Zinth, Bern Kohler, Thymine dimerization in DNA is an ultrafast photoreaction. Science **315**(5812), 625–629 (2007)
72. J.S. Taylor, Unraveling the molecular pathway from sunlight to skin cancer. Acc. Chem. Res. **27**(3), 76–82 (1994)
73. A.A. Vink, L. Roza, Biological consequences of cyclobutane pyrimidine dimers. J. Photochem. Photobiol. B: Biol. **65**(2), 101–104 (2001)
74. V.O. Melnikova, H.N. Ananthaswamy, Cellular and molecular events leading to the development of skin cancer. Mutat. Res./Fundam. Mol. Mech. Mutagen. **571**(1), 91–106 (2005). Biological Effects of Ultraviolet Radiation
75. D.L. Mitchell, R.S. Nairn, The biology of the (6-4) photoproduct. Photochem. Photobiol. **49**(6), 805–819 (1989)
76. J. Lim, T. Yatsui, M. Ohtsu, Observation of size-dependent resonance of near-field coupling between a deposited Zn dot and the probe apex during near-field optical chemical vapor deposition. IEICE Trans. Electron. **E88-C**(9), 832–1835 (2005)

Chapter 5
Wave Vector Excitation

5.1 Introduction

Si is widely used in most of the electric devices; therefore, our world cannot be without Si. To improve the performance of Si devices using electron only as carriers, it is required to reduce the size of the device below 10 nm. In addition to the electrons, it is expected to improve further by introducing the photons [1]. However, Si is not good material for photonic device because Si has low efficiency to excite the carrier by the photo-excitation of propagating FF. This is due to its indirect bandgap, in which Si has different wave vector between the Γ point and the X point [2, 3] (refer to Fig. 2.8a). Therefore, phonon assistance is required to compensate for the wave vector difference. To realize higher efficiency of photo-excitation, the usage of the plasmon resonance is widely used to enhance the optical-field intensity [4–6]. In addition to the plasmon resonance, further enhancement of the light intensity is realized by introducing three-dimensional structures to focus the light of the incident light to the detection area [7]. Although such techniques resulted in improvement of the absorption efficiency, it is still required phonon assistance. As discussed in Chap. 4, the ONF confined in a nanoscale could generate large components of the wave vector [8–10]; therefore, it is expected the usage of the ONF realizes a direct excitation of Si without phonon assistance. In this chapter, to evaluate the effect of ONF in the Si, we prepared the Si photo-detector with nanoscale structures as the ONF sources [11, 12].

5.2 Si p-n Photodiode

To utilize the ONF in photo-excitation process, we used the lateral p-n junction. This was because of the nano-scale size of ONF. Therefore, we need to fabricate the lateral p-n junction of the photodiode so that the p-n junction is close to the source of the ONF. We used an n-type Si, and the device was fabricated as following steps:

T. Yatsui, *Nanophotonic Chemical Reactions*, Nano-Optics and Nanophotonics,
https://doi.org/10.1007/978-3-030-42843-3_5

(i) The photomask was fabricated by the use of electron beam (E-beam) lithography.
(ii) The initial dopant density of Si substrate was 10^{15} cm^{-3}. The substrate was cleaned with a piranha and a hydrogen fluoride (HF) solution.
(iii) As for the mask in patterning device, the SiO_2 layer with 100 nm was prepared using thermal oxidization.
(iv) We used a photo-resist (ZPN, ZEON Co.) to pattern the SiO_2 layer in photolithography process.
(v) To produce p-type layer, the Spin on Dopant (SOD) (B153, Filmtronics) of the p-type dopant was coated (3,000 rpm for spin coating with 10 s), and then the coated SOD was baked (10 min at 100 °C) on a hot plate.
(vil) We used a furnace for boron doping (1000 °C for 60 min with O_2 (0.5 L min^{-1}) and N_2 (1.0 L min^{-1})), resulting 10^{20} cm^{-3} with a thickness of 0.9 μm of p-layer (Fig. 5.1a) [13].
(vil) We used the HF solution to remove the SiO_2 layer.
(vili) We fabricated the electrode using the lift-off technique. The electrode consisted of a 200-nm-thick Au with the thin Cr layer. We used acetone with ultrasonic cleaning for the removal of the resist used for the lift-off process.

By taking into account the dopant densities of the p- and n-type Si, the width of the depletion layer in the n-layer W was 1.06 μm. We fabricated the lateral p-n junction in comb shape to increase the area of the depletion layer and resultant larger carriers' density (Fig. 5.1b, c). As for the ONF source, we used Au nanoparticles (NPs, Tanaka Kikinzoku Kogyo). They were dispersed on the device (Fig. 5.2a). The solution including the Au NPs was dropped on the Si device, and the device was heated at 75 °C for 1 min to evaporate the solvent. We repeated this procedure N times. Figure 5.2b shows the Au NPs coverage depending on the values of N.

We measured the photo-sensitivity of the thick Si photo-detector with and without Au NPs to evaluate the ONF effect. We used a halogen lamp for the evaluation of wavelength dependence of the photo-sensitivity. We obtained the photo-sensitivity [A W^{-1}] using the values of the photo-current [A] and the incident power [W]. First, we measured the bare photo-sensitivity, i.e., without the Au NPs. Then, Au NPs were dropped on the device with dispersion number N. Finally, we measured the photo-sensitivity to evaluate the ONF effect. We measured at the same position between without and with Au NPs. Figure 5.3a is the photo-sensitivity of the thick Si device (thickness of Si substrate was 525 μm, see Fig. 5.1a). The solid red circles in Fig. 5.3a show the thick device with 100-nm Au NPs ($N = 5$). By comparing the photo-sensitivity, we evaluated the increased photo-sensitivity rate (Fig. 5.3b). The increased rate was the normalized photo-sensitivity with Au NPs by that without Au NPs. We found that the higher increased rate for larger N. The highest increased rate was 47.0% at $\lambda = 1100$ nm and $N = 10$. Note that the increased photo-sensitivity rate increased as the wavelength increased. The Au NP with 100-nm diameter in air has an absorption peak with a wavelength of 520 nm owing to the plasmon resonance [14]. The plasmon peak could change owing to the substrate.

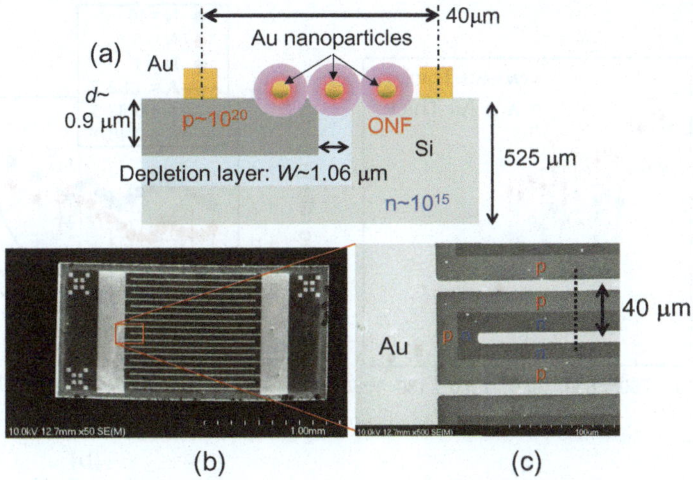

Fig. 5.1 **a** Schematic of the thick (525 μm) Si photo-detector. The thickness of the depletion layer W was 1.06 μm. d indicates the thickness of p-layer. SEM images of **b** the Si detector and **c** the magnified one of **b**. These figures are reprinted from [11] under the terms of the creative commons attribution license

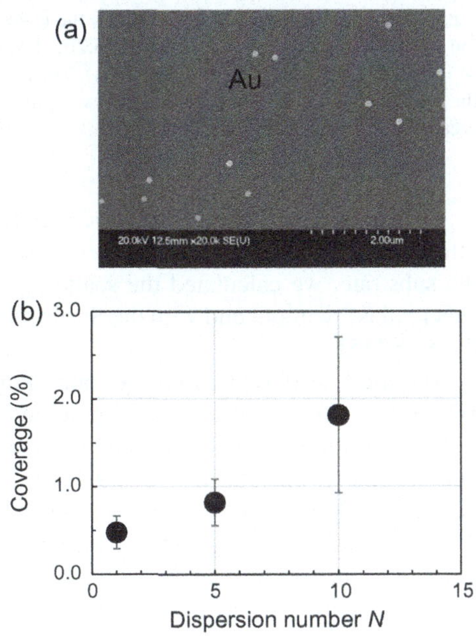

Fig. 5.2 **a** SEM image of the Au NPs ($N = 5$, D (diameter of Au) = 100 nm). **b** Dispersion number N dependence of the Au NPs ($D = 100$ nm) coverage. The error bar was obtained as the standard deviation of the coverage. The figure is reprinted from [11] under the terms of the creative commons attribution license

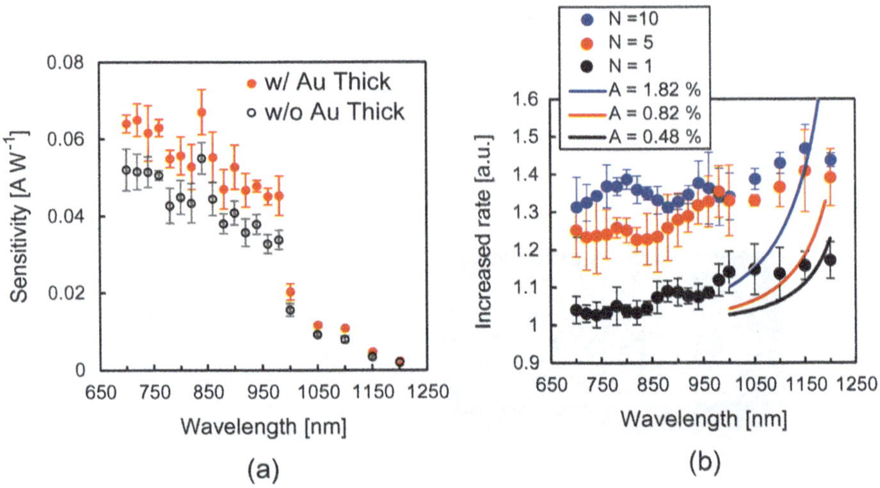

Fig. 5.3 **a** Excitation wavelength dependence of photo-sensitivity of thick Si device. The solid red circles and the open black circles are the photo-sensitivity of thick Si device with 100-nm Au NPs ($N = 5$) and without Au NPs, respectively. The error bar corresponds to the standard deviation of the photo-sensitivity for three samples. **b** Excitation wavelength dependence of increased photo-sensitivity rate for different N (=1, 5, and 10). The solid curves represent the increased photo-sensitivity rate calculated using (5.5). $A$is the ratios of the direct transition, and it was determined the Au NPs coverage as shown in Fig. 5.2b: A was set to 1.82% for $N = 10$ (blue), 0.82% for $N = 5$ (red), and 0.48% for $N = 1$ (black). The error bar was obtained as the standard deviation of the increased rate for three samples. These figures are reprinted from [11] under the terms of the creative commons attribution license

The absorption spectra might be shifted due to other possibilities, such as the effect of the substrate. To evaluate the spectral redshift in the absorbance due to the substrate, we calculated the scattering cross section (SCS) of the 100-nm Au spherical NP without and with the Si substrate. We used the FDTD method to obtain the SCS [15].

The used an FDTD-based program "Poynting for Optics" (product of Fujitsu). We set boundary conditions as a periodic boundary condition (PBC) in the x- and y-directions and perfectly matched layer (PML) in the z-direction. The basic computational area was 600 × 600 × 9,500 (–5,000 nm $\leq z \leq$ 4,500) nm (see Fig. 5.4). As for the 100-nm Au NP, the area coverage of NP over a 600 × 600 nm surface was 2.2%. This value is comparable to the experimental one (~2%). All light sources had a uniform electric field intensity distribution with an electric field intensity of 1.0 [V m^{-1}] in the xy-plane at $z = 4{,}200$ nm. The incident light had a x-polarized light. The minimum and maximum cell sizes were 0.22 × 0.22 × 0.48 nm and 10 × 10 × 10 nm, respectively. Note that we used same unit cells for all calculation regardless of different Au NPs size without and with the Si substrate. The refractive indexes were $0.31 + i\ 7.93$ (Au) and 3.54 (Si), respectively. Because we observed the larger increased photo-sensitivity $\lambda > 1{,}000$ nm, we calculated at $\lambda = 1{,}100$ nm.

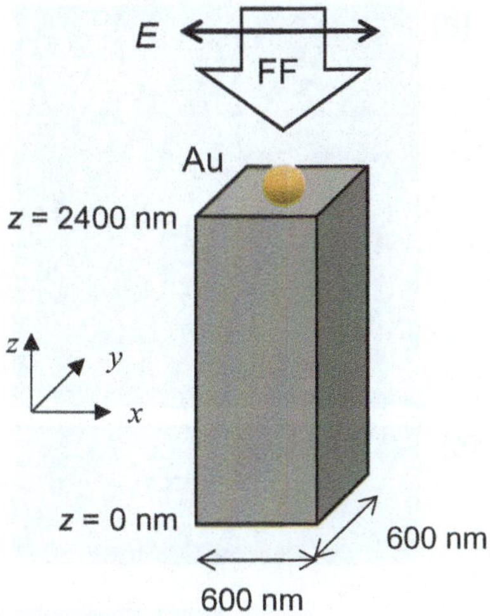

Fig. 5.4 Schematic of the calculation model for FDTD. FF represents far-field. The figure is reprinted from [11] under the terms of the creative commons attribution license

To evaluate the NF effect, we obtained the electric field intensity distribution at the Au NP/Si substrate interface. Figure 5.5a–e shows the intensity of electric field distributions of Au NPs with $D = 100\,\text{nm}$.

We also calculated the Poynting vectors. The white arrows (see Fig. 5.5e) show the calculated Poynting vector. From the results, the electric field had the flows vertically to the Si substrate, and not horizontally. The results support that the effect of the scattering enhancement from the Au NP was negligible. In addition, the light scattering effect should be excluded because the scattering intensity depended on λ^{-4} [16, 17].

Next, we calculated the SCS of the spherical Au NP ($D = 100\,\text{nm}$) without and with the Si substrate ($600 \times 600 \times 2{,}400\,\text{nm}$. See Fig. 5.4). To obtain the SCS, we used a light source with a differential Gaussian wave in the time domain, in which a wavelength calculation range was between 400 and 1,300 nm. As a reference, the fields without the Au NP was calculated. A light source had an uniform electric field intensity in the xy-plane. The angle resolution was to 5° [18].

The solid lines in Fig. 5.6 are the normalized SCS, implying that the Au NP spectra both without and with the Si substrate had highest value at $\lambda = 520\,\text{nm}$ owing to the plasmon resonance. The intensity of the SCS with the Si substrate (solid red line, Fig. 5.6) was stronger than that without the Si substrate (solid blue line, Fig. 5.6). Note that there is no difference of peak wavelength between without and with the Si substrate. The reason of intensity enhancement putting on the substrate was considered to the mirror image effect. To confirm this postulate, we obtained analytical solution by calculating the SCS using the dipole approximation [19]. We can obtain the polarizability of the spherical Au with the Si substrate using the following equation:

$$\chi = \frac{\varepsilon_{\text{Air}}\left(\varepsilon_{\text{Au}} - \varepsilon_{\text{Air}}\right) V}{\varepsilon_{\text{Air}} + L\left(\varepsilon_{\text{Au}} - \varepsilon_{\text{Air}}\right)} \tag{5.1}$$

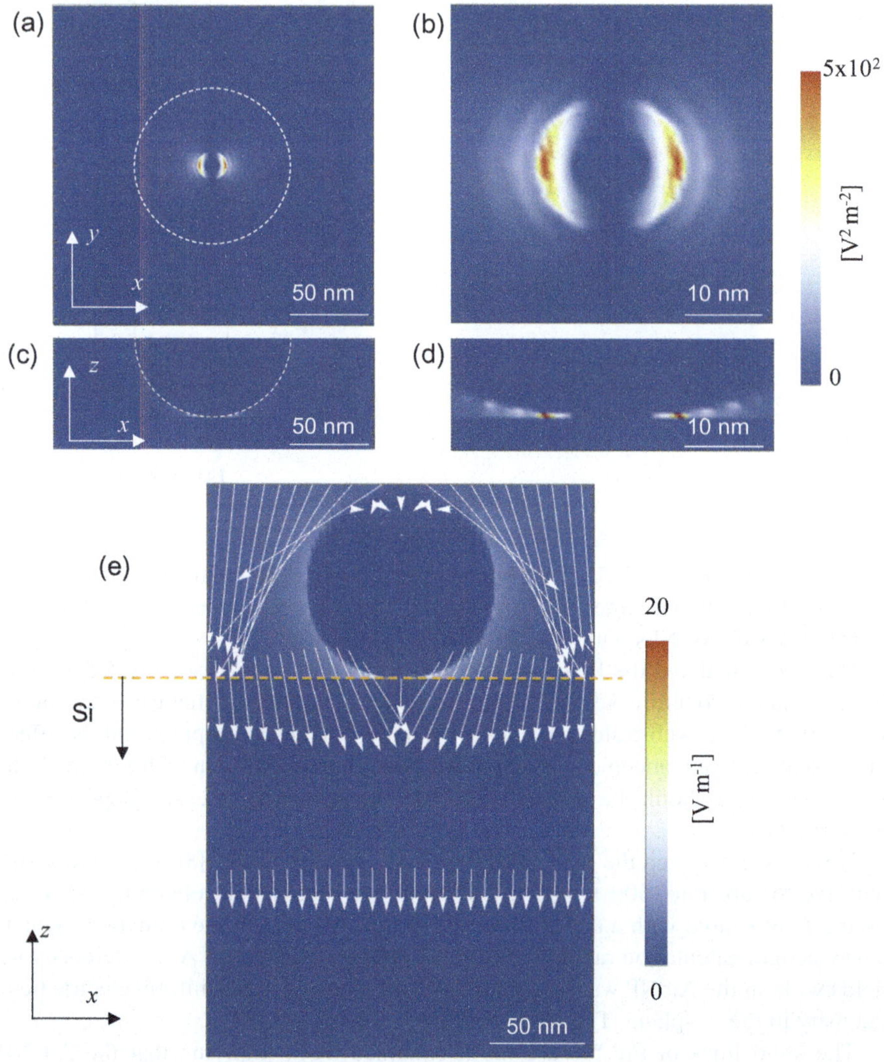

Fig. 5.5 **a–d** Electric field intensity distribution E^2 at the Au NP ($D = 100\,\mathrm{nm}$)/Si substrate. The dashed white circles in **a** and **c** represent the outline of Au NP. **e** Poynting vector distributions. Electric field distribution |E| at the 100-nm Au NP/2.4 μm Si interface. Yellow dashed line represents the top surface of Si. White arrows represent the Poynting vector. The figures are reprinted from [11] under the terms of the creative commons attribution license

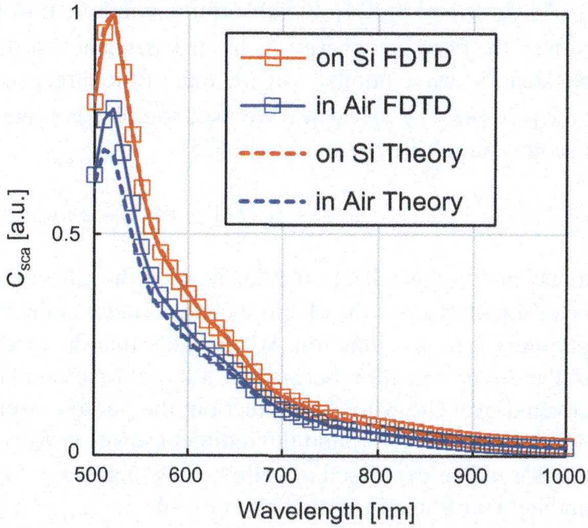

Fig. 5.6 Calculated SCS C_{sca}. The solid lines and open squares are the FDTD results with Si substrate (red) and without Si substrate (blue), respectively. The dashed lines are the analytical solutions using (5.5) with Si substrate (red) and without Si substrate (blue). The figure is reprinted from [11] under the terms of the creative commons attribution license

where ε_{Au} and ε_{Air} are the permittivity of gold and air, respectively, and V is the volume of the Au sphere. L is the parameter defined using the following equation:

$$L = \frac{1}{3}\left(1 - \frac{1}{8}\frac{\varepsilon_{Si} - \varepsilon_{Air}}{\varepsilon_{Si} + \varepsilon_{Air}}\right) \tag{5.2}$$

where ε_{Si} is the permittivity of Si. We set the virtual dipole in the Si substrate, in which Si surface was plane of mirror symmetry. The SCS is calculated as:

$$C_{sca} = \frac{k^4}{6\pi}\left|\chi^2\right| \tag{5.3}$$

in which k and χ represent the wave number and the value obtained using (5.1). We obtained the SCS (dashed lines, Fig. 5.6) obtained by (5.3). Note that the FDTD spectra were good agreement with the that obtained by analytical solution (5.3). The results support the above postulate that the intensity increase of Au NP with the Si substrate owing to the mirror image effect. One of the possibilities of the redshift is the ellipsoid shape [20, 21]. However, this possibility can be excluded, because our Au NP had a shape close to a true sphere, not an oval sphere. As can be seen in Fig. 5.2b, the area coverage of the Au NPs was $\sim$1.8%. Therefore, higher increased photo-sensitivity rate could be realized by the increase in area coverage of Au NPs.

Most of the reports on the increase in the sensitivity with metallic NPs utilized high metallic area coverage more than 50% [20, 22, 23], in which the increased photo-sensitivity was obtained by a longer optical pass in the detector due to the scattering from metallic NP [20, 22, 23].

As described in [24], the absorption coefficient of the indirect bandgap material α_I near the band gap energy E_g has a dependence of $\alpha_I \propto \left(h\nu - E_g\right)^2 / h\nu$ (h: Plank constant, ν: wave number), while that of the direct bandgap material α_D depended on $\alpha_D \propto \left(h\nu - E_g\right)^{1/2} / h\nu$. We used the bias voltage of 0; then, the photo-current becomes the short circuit current [25]:

$$I_{SC} = Q\,(1 - r)\,\{1 - \exp(-\alpha l)\}\,en \tag{5.4}$$

in which Q is the collection efficiency, r is the reflection coefficient, l is the absorbing layer thickness, e is the electron charge, and n is the number of photons per second per unit of the p-n junction. We consider that the ONF generated at Au NPs results in the direct transition because of a large wave number. Because Au NPs were not covered over the whole p-n junction, the photo-current had the current both of the direct transition I_{SC_D} and the indirect transition I_{SC_I}. The increased photo-current rate should be expressed as follows in which we used respective ratios of the direct and indirect transitions as A and $(1 - A)$:

$$\begin{aligned}\frac{AI_{SC\_D} + (1 - A)\,I_{SC\_I}}{I_{SC\_I}} &= \frac{A\,(1 - \exp(-\alpha_D l)) + (1 - A)\,(1 - \exp(-\alpha_I l))}{1 - \exp(-\alpha_I l)} \\ &\approx \frac{(A\alpha_D + (1 - A)\,\alpha_I)\,l}{\alpha_I l} \\ &= C\frac{A\left(h\nu - E_g\right)^{1/2} + (1 - A)\left(h\nu - E_g\right)^2}{\left(h\nu - E_g\right)^2}\end{aligned} \tag{5.5}$$

in which we calculated using the Taylor series of the exponential function, and C represents the proportional constant. The solid curves in Fig. 5.3b show the calculated increased rate using (5.5). In the simulation, we used $E_g = 0.96\,\text{eV}$, which was an average value of $N = 1$, 5, and 10 obtained from fitting by the relation $I_{SC\_I} \propto \alpha_I \propto \left(h\nu - E_g\right)^2 / h\nu$ for the photo-sensitivity data (Fig. 5.3a). Although the obtained value ($E_g = 0.96\,\text{eV}$) for the band gap energy of Si is smaller than the band gap of bulk Si (1.1 eV), heavy doping, as high as $10^{20}\,\text{cm}^{-3}$, results in the band narrowing, i.e., decrease in the band gap [26]. We calculated wavelength range from 1,000 to 1,200 nm, because we took into consideration the behavior near the wavelength of band gap energy. In addition, we obtained the wavelength dependence of the increased rate between 700 and 1,200 nm. The respective ratios of the direct transition A were 0.48 ($N = 1$), 0.82 ($N = 5$), and 1.82% ($N = 10$) for the average area coverage. From the comparison of the increased rate (see Fig. 5.3b), the calculated increased rates did not fit well with experimental ones. We considered that this difference was caused by the narrow depletion layer on the Si surface as small as 1.06 μm. In this experiment, the Au NPs were randomly deposited on the whole area, including p-type, n-type, and depletion layers. The width of the depletion layer (1.06 μm) was ~1/40 of the electrode separation (40 μm). However, we observed the higher increased rate as the wavelength increases to the wavelength of the band gap energy from both calculated and experimental data supported the direct transition by the ONF in the Si photo-detector.

We obtained increased photo-sensitivity by using the Au NPs. In our device, the penetration depth of the Si near the band gap wavelength is estimated $\geq 10\,\mu m$, which is longer than $0.9\,\mu m$ (the p-layer depth). Therefore, most of the generated carriers in the thick Si photo-detector should be originated from the Si substrate beneath the p-layer, i.e., the increased photo-sensitivity might be originated from the FF effect. To exclude the FF excitation, we fabricated a thin ($0.9\,\mu m$) Si photo-detector using an n-type silicon-on-insulator (SOI) with a dopant density of $10^{13}\,cm^{-3}$ (Fig. 5.7a). The buried oxide (BOX) layer thickness in the SOI wafer was $1.5\,\mu m$. The p-type layer with the dopant density of $10^{20}\,cm^{-3}$ was obtained by the thermal diffusion of boron. Taking into account the dopant densities of the p- and n-type Si, the depletion layer thickness w was $\sim 9.92\,\mu m$ in the n-type Si only. The Au NPs were dispersed on the photo-detector region after device fabrication process.

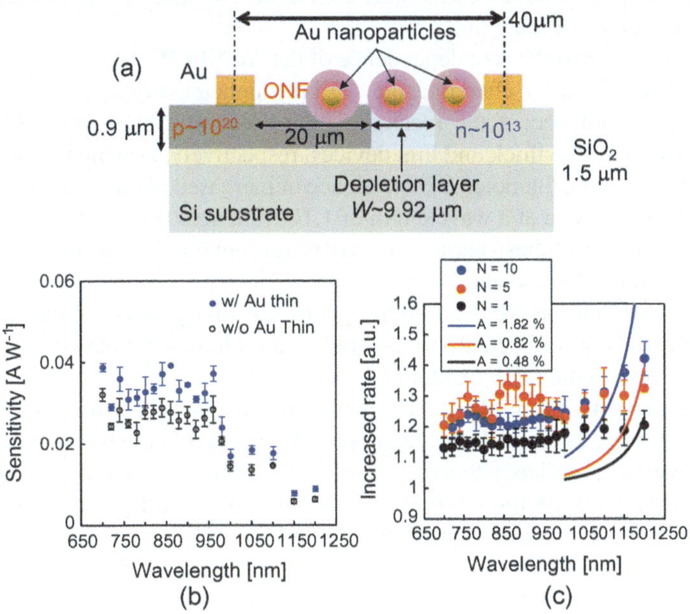

Fig. 5.7 **a** Schematic of the thin device using a silicon-on-insulator (SOI). W represents the depletion layer thickness. **b** The excitation wavelength dependence of sensitivity of the thin Si devices. The solid blue circles and the open black circles are the thin device with the 100-nm Au NPs ($N = 5$) and the bare thin device, i.e., no Au NP. The error bar corresponds to the standard deviation of the sensitivity for three samples. **c** The excitation wavelength dependence of the increased photo-sensitivity rate. The solid curves are calculated by increased rate obtained from (5.5), in which the ratios of the direct transition A were set to 1.82% for $N = 10$ (blue), 0.82% for $N = 5$ (red), and 0.48% for $N = 1$ (black). The error bar corresponds to the standard deviation of the increased rate for three samples. These figures are reprinted from [11] under the terms of the creative commons attribution license

First, we compared the photo-sensitivity properties between the thick and thin devices. Figure 5.7b is the wavelength dependence of photo-sensitivity of thin Si device. We obtained three samples for each device. The solid blue circles in Fig. 5.7b are the thin device with $N = 5$, of the 100-nm Au NPs. We found that the thin device has two times lower photo-sensitivity than that of the thick device, implying no the FF excitation effect in the photo-sensitivity of thin device. Figure 5.7c is the wavelength dependence of increased photo-sensitivity rate for different N (=1, 5, and 10). Note that we found higher increased photo-sensitivity rate as wavelength increased. The solid curves in Fig. 5.7c are the calculated increased rate obtained from (5.5). We found that the gradient of the increased rate of $N = 10$ in the wavelength range from 1,000 to 1,200 nm was 0.0009 nm^{-1}, which was 1.8 times larger than that for the thick one of $N = 10$ in the same wavelength range. We obtained highest increased rate of 42.3% N at $\lambda = 1{,}200$ nm with $N = 10$. As mentioned above, the FF excitation was eliminated from thin device, we considered that the increase in the photo-sensitivity rate was resulted from the NF excitation. It is expected that larger value of photo-sensitivity could be obtained by a higher Au NP density and a wider depletion layer, for example of a p-i-n junction.

Next, we evaluated the size dependence of the Au NPs We obtained the increased photo-sensitivity rate by changing the Au NP diameter. Figure 5.8a, b shows the wavelength dependence of the increased photo-sensitivity rate with different Au NP diameter D for the thick and thin devices, respectively, in which N was set to 5. Figure 5.8c shows the diameter D dependence of increased photo-sensitivity rate. All experiments were done at a wavelength of 1,100 nm, near the band gap wavelength. It is noted that the highest photo-sensitivity was obtained with an Au NP with a $D = 100$ nm for both devices.

It was found that Au NPs had higher photo-sensitivity as D increased. The size dependence of the photo-sensitivity contradicted with the postulate of direct optical transition owing to the ONF excitation.

In the uncertainty principle of the direct optical transition by the ONF, we should obtain a larger Δk as the NP size decreases. We calculated the NP size dependence to clarify the NP size dependence. To evaluate the NF component, we put Au NPs on the Si substrate (600 nm $\times$ 600 nm $\times$ 2400 nm, same model as the case shown in Fig. 5.4). We obtained similar distributions for the Au NP on the SOI (Si: 600 nm $\times$ 600 nm $\times$ 900 nm on SiO: 600 nm $\times$ 600 nm $\times$ 1500 nm). Therefore, we considered the distribution only on the Si substrate. The unit cell was set to be the same for all the Au NP sizes. We used the refractive indexes of $0.31 + i\ 7.93$ (Au) and 3.54 (Si), respectively. We calculated using continuous wave (CW) light source with λ of 1,100 nm. We used a linearly polarized light with x-polarization in the calculation. To evaluate the NF effect, we calculated the electric field intensity distribution at the Au NP/Si substrate interface (Fig. 5.5a–d). Then, we calculated the wave number using the Fourier transform of the electric field distribution along the x-axis.

Figure 5.9a is Au NP sizes (D) dependence of an averaged power spectrum. The spectra were averaged in the range of $0 \leq y \leq D/2$. The difference of wave number between Γ-X points is $k_{x\_\Gamma X} = 0.85 \times \pi/a$ [27]. Using $a = 5.43 \times 10^{-1}$ nm for Si, $k_{x\_\Gamma X}$ is obtained as 4.92 nm^{-1}. The red solid circles in Fig. 5.9b is the intensity of

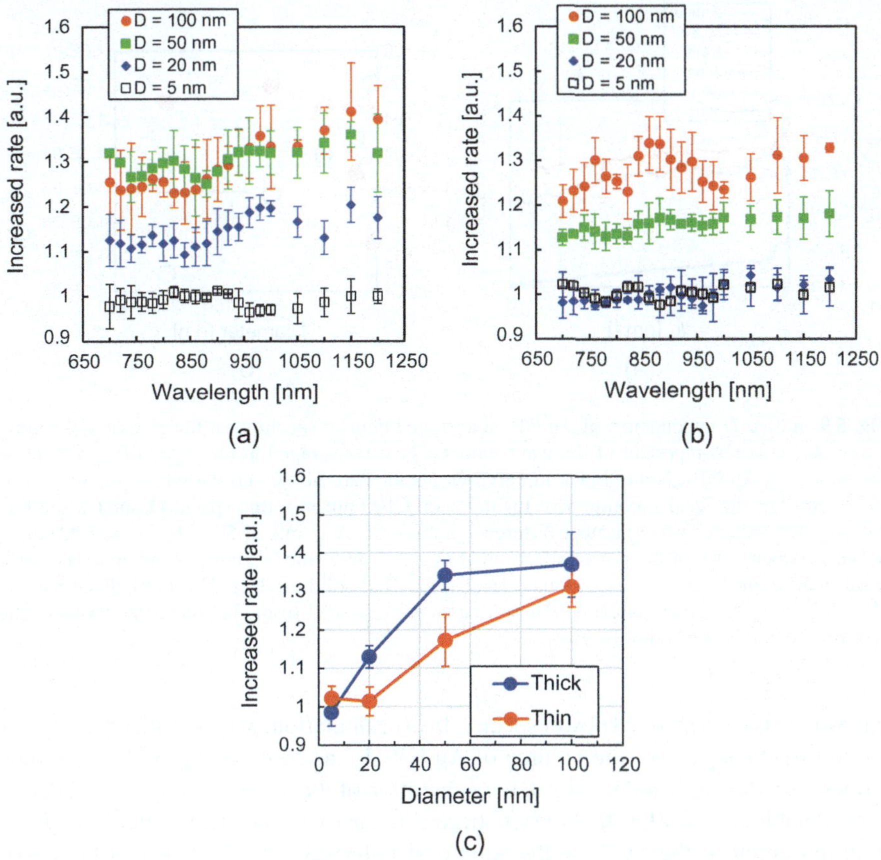

Fig. 5.8 Wavelength dependence of the increased photo-sensitivity for **a** thick and **b** thin devices. *D* represents the diameter of Au NP. The error bar corresponds to the standard deviation of the increased rate for three samples. **c** Diameter *D* dependence of increased photo-sensitivity rate obtained at $\lambda = 1{,}100\,\text{nm}$. The error bar corresponds to the standard deviation of the increased rate for three samples. These figures are reprinted from [11] under the terms of the creative commons attribution license

power spectra at $k_{x\_\Gamma X} = 4.92\,\text{nm}^{-1}$ in Fig. 5.8a ($|F(E)_{\Gamma X}|^2$). From the results, we obtained the highest intensity in the case of $D = 100\,\text{nm}$, which is exceeding 10,000 times as much as that in the case of $D = 5\,\text{nm}$. From the results, the calculated size dependence, taking the maximum at $D = 100\,\text{nm}$ (Fig. 5.8b), was good agree with that of experimental results (Fig. 5.8c). The electric field strongly localized in the Au NP/Si interface as can be seen in Fig. 5.5a–d. Therefore, the decrease in the value of $|F(E)_{\Gamma X}|^2$ at $D = 200\,\text{nm}$ should be originated from the absorption in Au. However, the calculated results contradict the postulate of direct optical transition by the non-uniform optical field owing to the uncertainty principle. To estimate the effect of k-component generation by the non-uniform optical field, we obtained the

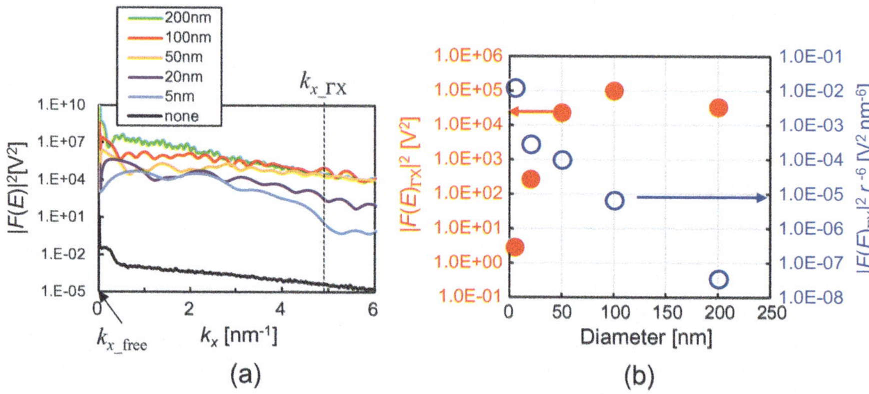

Fig. 5.9 **a** Size D dependence of Au NP on averaged Fourier spectrum of the electric field along x-axis. k_x: x-axis component of the wave number. k_x was averaged in the rage of $0 \leq y \leq D/2$. D: the size of Au NPs. None (black line) represents the bare Si, i.e., no Au NP. k_{x_free} (=5.71 × 10^{-3} [nm^{-1}]): the ideal k-component for the λ of 1,100 nm in a free space. Dashed white line indicates the value of wave number difference between Γ-X points of Si of $k_{x\_\Gamma X} = 4.92\,\text{nm}^{-1}$. **b** The D dependence of the power spectrum at $k_{x\_\Gamma X} = 4.92$ [nm^{-1}], corresponding to the wave number difference between Γ-X points: $|F(E)_{\Gamma X}|^2$ (red solid circles). The normalized by V^2: $|F(E)_{\Gamma X}|^2 r^{-6}$ (blue open circles). These figures are reprinted from [11] under the terms of the creative commons attribution license

intensity of the normalized power spectra. In the calculation, we normalized the power spectra by the square of the volume of Au NPs V^2 as show in Fig. 5.8b (open blue circles). Scattering at NP results in the generation of the large k-component. Then, it is reasonable to consider the Fourier spectral intensity of the electric field, $|F(E)|^2$, is proportional to the SCS ($\propto$ the square of polarizability [28]). It is known that polarizability is proportional to the volume V [29], it is reasonable of this procedure in which the value $|F(E)|^2$ was normalize by r^6 ($\propto V^2$) [30]. Then, we obtained the size dependence that the intensity of the normalized power spectra decreased as the size increase. From the calculated results using FDTD, it was confirmed that a greater Δk is generated as the size decreases, i.e., the direct optical transition efficiency increased by introducing the ONF. Furthermore, by considering the V, the photo-sensitivity increases as the size increases.

References

1. M. Paniccia, Silicon photonics is this the future? in *Nature Photonics Conference* (2007)
2. H. Lu, G. Chen, Analysis of optical absorption in silicon nanowire arrays for photovoltaic applications. Nano Lett. **7**(11), 3249–3252 (2007)
3. J. Noffsinger, E. Kioupakis, C.G. Van de Walle, S.G. Louie, M.L. Cohen, Phonon-assisted optical absorption in silicon from first principles. Phys. Rev. Lett. **108**, 167402 (2012)
4. D.M. Schaadt, B. Feng, E.T. Yu, Enhanced semiconductor optical absorption via surface plasmon excitation in metal nanoparticles. Appl. Phys. Lett. **86**(6), 063106 (2005)

5. H.A. Atwater, A. Polman, Plasmonics for improved photovoltaic devices. Nat. Mater. **9**, 205–213 (2010)
6. I. Goykhman, B. Desiatov, J. Khurgin, J. Shappir, U. Levy, Locally oxidized silicon surface-plasmon Schottky detector for telecom regime. Nano Lett. **11**(6), 2219–2224 (2011)
7. M.D. Kelzenberg, S.W. Boettcher, J.A. Petykiewicz, D.B. Turner-Evans, M.C. Putnam, E.L. Warren, J.M. Spurgeon, R.M. Briggs, N.S. Lewis, H.A. Atwater, Enhanced absorption and carrier collection in Si wire arrays for photovoltaic applications. Nat. Mater. **9**, 239–244 (2010)
8. T. Yatsui, M. Yamaguchi, K. Nobusada, Nano-scale chemical reactions based on non-uniform optical near-fields and their applications. Prog. Quantum Electron. **55**, 166–194 (2017)
9. J. Jung, M.L. Trolle, K. Pedersen, T.G. Pedersen, Indirect near-field absorption mediated by localized surface plasmons. Phys. Rev. B **84**, 165447 (2011)
10. M. Yamaguchi, K. Nobusada, Indirect interband transition induced by optical near fields with large wave numbers. Phys. Rev. B **93**, 195111 (2016)
11. T. Yatsui, S. Okada, T. Takemori, T.S.K. Saichi, T. Ogamoto, S. Chiashi, S. Maruyama, M. Noda, K. Yabana, K. Iida, K. Nobusada, Enhanced photo-sensitivity in a Si photodetector using a near-field assisted excitation. Commun. Phys. **2**, 62 (2019)
12. T. Yatsui, Recent improvement of silicon absorption in opto-electric devices. Opto-Electron. Adv. **2**(10), 190023 (2019)
13. J.C. Irvin, Resistivity of bulk silicon and of diffused layers in silicon. Bell Syst. Tech. J. **41**(2), 387–410 (1962)
14. S. Link, M.A. El-Sayed, Spectral properties and relaxation dynamics of surface plasmon electronic oscillations in gold and silver nanodots and nanorods. J. Phys. Chem. B **103**(40), 8410–8426 (1999)
15. V. Myroshnychenko, J. Rodriguez-Fernández, I. Pastoriza-Santos, A.M. Funston, C. Novo, P. Mulvaney, L.M. Liz-Marzán, F.J.G. de Abajo, Modelling the optical response of gold nanoparticles. Chem. Soc. Rev. **37**, 1792–1805 (2008)
16. L. Rayleigh F.R.S., XXXIV. On the transmission of light through an atmosphere containing small particles in suspension, and on the origin of the blue of the sky. Lond. Edinb. Dublin Philos. Mag. J. Sci. **47**(287), 375–384 (1899)
17. A.T. Young, Rayleigh scattering. Phys. Today **35**(1), 42–48 (1982)
18. A. Taflove, S.C. Hagness, *Computational Electrodynamics: The Finite-Difference Time-Domain Method* (Artech House, Boston, 2005)
19. M.M. Wind, J. Vlieger, D. Bedeaux, The polarizability of a truncated sphere on a substrate I. Phys. A: Stat. Mech. Appl. **141**(1), 33–57 (1987)
20. K. Nakayama, K. Tanabe, H.A. Atwater, Plasmonic nanoparticle enhanced light absorption in GaAs solar cells. Appl. Phys. Lett. **93**(12), 121904 (2008)
21. K.R. Catchpole, A. Polman, Design principles for particle plasmon enhanced solar cells. Appl. Phys. Lett. **93**(19), 191113 (2008)
22. S. Pillai, K.R. Catchpole, T. Trupke, M.A. Green, Surface plasmon enhanced silicon solar cells. J. Appl. Phys. **101**(9), 093105 (2007)
23. H.R. Stuart, D.G. Hall, Absorption enhancement in silicon-on-insulator waveguides using metal island films. Appl. Phys. Lett. **69**(16), 2327–2329 (1996)
24. J.I. Pankove, *Optical Processes in Semiconductors* (Dover, New York, 1971)
25. J.J. Loferski, Theoretical considerations governing the choice of the optimum semiconductor for photovoltaic solar energy conversion. J. Appl. Phys. **27**(7), 777–784 (1956)
26. J. Wagner, J.A. del Alamo, Band-gap narrowing in heavily doped silicon: a comparison of optical and electrical data. J. Appl. Phys. **63**(2), 425–429 (1988)
27. S. Richard, F. Aniel, G. Fishman, Energy-band structure of Ge, Si, and GaAs: a thirty-band $\mathbf{k}\cdot\mathbf{p}$ method. Phys. Rev. B **70**, 235204 (2004)
28. C.F. Bohren, D.R. Huffman, *Absorption and Scattering of Light by Small Particles* (Wiley, New York, 1998)
29. J.D. Jackson, *Classical Electrodynamics* (Wiley, New York, 1962)
30. S.A. Maier, *Plasmonics: Fundamentals and Applications* (Springer, Berlin, 2007)

Chapter 6
Remarks

The previous sections have described how the non-uniform optical field (ONF) is used. The phenomena of the ONF are unique, not just about reducing their size. The uniqueness of the ONF could help in the realization of novel fabrication technique such as the near-field etching, which has not been achieved so far by any other technique. In addition, since the ONF makes it possible to utilize wavelengths longer than absorption wavelength, we could utilize the wavelengths which are not conventionally used. Therefore, we expect that the use of the ONF could accelerate realization of some of the sustainable development goals (SDGs) [1]. In order to utilize these phenomena in the practical applications, they might need to combine with other phenomena used in plasmonics and metamaterial, together with other techniques such as self-assembly, to enhance the efficiency of photon use. Further, with the progress in numerical simulations, optimally designed structures become necessary to induce a non-uniform optical fields to match the applications.

It will be extremely rewarding if this book serves the purpose of teaching nanophotonics and aids in the readers' future study and research. In addition, for the further development and the spread of this technique, collaborations with the other fields are welcome. In particular, comments from the industry would be appreciated.

T. Yatsui, *Nanophotonic Chemical Reactions*, Nano-Optics and Nanophotonics,
https://doi.org/10.1007/978-3-030-42843-3_6

Reference

1. https://www.un.org/sustainabledevelopment/sustainable-development-goals/

Index

T. Yatsui, *Nanophotonic Chemical Reactions*, Nano-Optics and Nanophotonics,
https://doi.org/10.1007/978-3-030-42843-3

The manufacturer's authorised representative in the EU is Springer Nature Customer Service Centre GmbH, Europaplatz 3, 69115 Heidelberg, Germany. If you have any concerns regarding our products, please contact ProductSafety@springernature.com

Printed and bound by CPI Group (UK) Ltd, Croydon, CR0 4YY
09/09/2026
02213740-0001